APPLIED RESEARCH IN
FIELD CROP PATHOLOGY
FOR INDIANA, 2021

APPLIED RESEARCH IN FIELD CROP PATHOLOGY FOR INDIANA, 2021

DARCY E. P. TELENKO
AND SUJOUNG SHIM

PURDUE UNIVERSITY PRESS
WEST LAFAYETTE, INDIANA

Cataloging-in-Publication Data on file at the Library Congress.

ISBN 978-1-62671-286-7 (paperback)

ISBN 978-1-62671-287-4 (epdf)

CONTENTS

ACKNOWLEDGMENTS

This report is a summary of applied field crop pathology research trials conducted in 2021 under the direction of the Purdue Field Crop Pathology program in the Department of Botany and Plant Pathology at Purdue University. The authors wish to thank the Purdue Agronomy Research and Education Center, the Purdue Agricultural Centers, and the many cooperators and contributors who provided the resources needed to support the applied field crop pathology research program in Indiana. Special recognition is extended to Stephen Brand, Su Shim, and Camila Rocco da Silva for technical skills in managing field trials, data organization and processing, and help preparing this report; Mariama Brown, Tiffanna Ross, Audrey Conrad, and Kaitlin Waible, graduate students who assisted with field trial data collection and analysis; Emily Duncan, Audrey Toogood, and Autumn Greer, undergraduate student interns who assisted with field trial data collection and scouting; Dr. Tom Creswell, Dr. John Bonkowski, and Todd Abrahamson with the Purdue Plant Pest Diagnostic Laboratory for assistance in pathogen surveys and diagnosis; and Dr. Creswell and Dr. Bonkowski for providing peer review. Collectively, the contributions of colleagues, professionals, students, and growers were responsible for a highly successful and productive program to evaluate products and practices for disease management in field crops.

The authors would also like to thank the following for their support in 2021: Bayer Crop Science, BASF, Certis USA, Corteva Agriscience, FMC Agricultural Solution, Gowan, the Indiana Corn Marketing Council, the Indiana Soybean Alliance, North Central Soybean Research Program, Oro Agri, Pioneer, Purdue University, Sipcam Agro, Syngenta, UPD NA Inc., the USDA NIFA Hatch Project #1019253, USWBSI, and Valent.

SUMMARY OF 2021 FIELD CROP DISEASE SEASON

CORN

In 2021, there was moderate disease on corn in Indiana across the state; details of major issues are listed below. Gray leaf spot, northern corn leaf blight, northern corn leaf spot, and Diplodia streak were found in pockets. There were also numerous reports of Physoderma brown spot and stalk rot. Tar spot and southern rust were two diseases that were closely monitored this season.

Tar spot. Tar spot of corn was a concern in 2021 due to favorable weather conditions. In 2021, a widespread epidemic occurred in northern Indiana and in pockets in other areas of the state. The environmental conditions are key in determining field risk year to year, as leaf wetness plays an important role in tar spot disease development. The third year of tar spot–directed research has been completed here in Indiana. As a cautionary note, it is still important to have multiple years of data for verification, but the initial results do serve as a good starting point for making future management decisions.

The field crop pathology team made a large effort at the end of the season to scout for tar spot across the state. Four new counties were confirmed with tar spot in 2021, making 82 counties total in Indiana to date. Out of the 188 fields scouted, 143 were positive for tar spot (76.1%). In addition, incidence and severity were rated (examples of severity are in Figure 1) and used to generate a tar spot index, shown in the map in Figure 1, with increasing severity indicated by the darkness of the orange color of the county. The map demonstrates how corn produced in northern Indiana is at a higher risk for tar spot versus central and southern Indiana, but there are new pockets of disease emerging in Indiana. The map also parallels the weather conditions and reports during 2021. It is important to document tar spot movement in the state so that when favorable conditions arise, increased tar spot disease risk can be more accurately assessed across the remainder of the state.

Southern corn rust. Southern corn rust was first found in Indiana in the 2021 season on July 16, and by the end of the season a total of 73 counties were confirmed to have the disease present (Figure 2). Southern rust pustules generally tend to occur on the upper surface of the leaf and produce chlorotic symptoms on the underside of the leaf (Figure 2). These pustules rupture the leaf surface and are orange to tan in color. They are circular to oval in shape. Common rust was also widespread, and both diseases could be present on a leaf and easily mistaken for each other. It is important to send a sample to the Purdue Plant Pest Diagnostic Lab for confirmation if southern rust is suspected. There is an increased risk for yield impact if southern rust is identified early in the season.

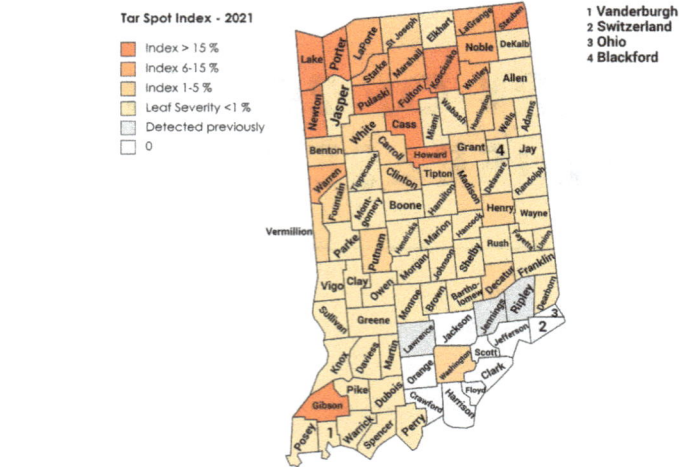

FIGURE 1. 2021 tar spot index for Indiana. The darker orange the county, the greater the field incidence and severity of tar spot in the fields in which it was found. The range of tar spot severity on leaves >25%, 5–7%, 1%, and <1%. Photo credit: D. Telenko.

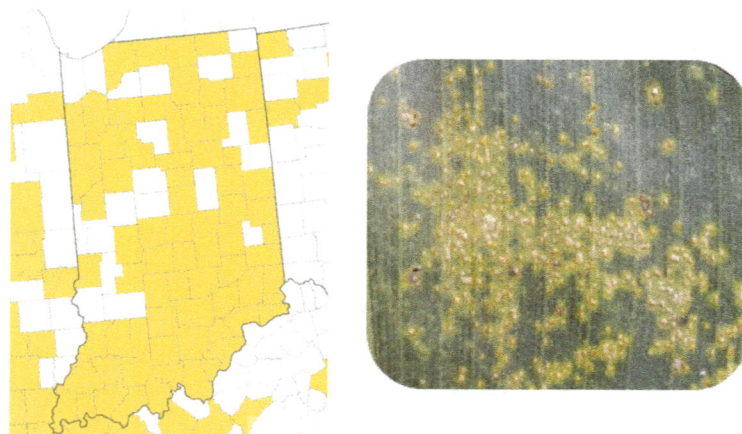

FIGURE 2. Southern corn rust map of confirmed (yellow) counties that had southern corn rust in Indiana in 2021 and a corn leaf with southern rust infection. Photo credit: D. Telenko, Map source https://corn.ipmpipe.org/southerncornrust/

Due to the need to monitor both southern rust and tar spot in Indiana, there will be **no charge for Indiana growers to submit southern rust and tar spot samples to the Plant Pest Diagnostic Lab for diagnostic confirmation.** This service is made possible through research supported by the Indiana Corn Marketing Council.

SOYBEAN

Diseases in soybeans remained relatively low throughout the season for much of the state. Our research sites and sentinel plots across the state saw low levels of frogeye leaf spot, Cercospora leaf blight, downy mildew, and Septoria brown spot. There were pockets where sudden death syndrome and white mold caused issues in fields. In general, it was a quiet year for foliar diseases in soybean.

WHEAT

Fusarium head blight (FHB), or scab, is one of the most impactful diseases of wheat and among most the challenging to prevent. In addition, FHB infection can cause the production of a mycotoxin called deoxynivalenol (DON), or vomitoxin. The conditions in 2021 were less conducive to FHB development. Our research sites in both West Lafayette and Vincennes had low levels of FHB develop in our nontreated susceptible cultivar checks, and initial DON testing was less than 1 ppm. FHB management requires an integrated approach, including the selection of varieties with moderate resistance and timely fungicide application at flowering. Other diseases observed in our wheat trials in 2021 included leaf rust and Septoria leaf and glume blotch.

AGRONOMY CENTER FOR RESEARCH AND EDUCATION (ACRE)

EVALUATION OF FUNGICIDES FOR FOLIAR DISEASES IN CORN IN CENTRAL INDIANA, 2021 (COR21-01.ACRE)

A. L. Greer, S. Shim, and D. E. P. Telenko, Department of Botany and Plant Pathology, Purdue University West Lafayette, IN 47907-2054

CORN (*ZEA MAYS* P0574AMXT)

Gray leaf spot, *Cercospora zeae-maydis*
Tar spot, *Phyllachora maydis*

A trial was established at the Purdue Agronomy Center for Research and Education (ACRE) in Tippecanoe County, Indiana. The trial was a randomized complete block design with four replications. Plots were 10 feet wide and 30 feet long and consisted of four rows, with the two center rows were used for evaluation. The previous crop was corn. Standard practices for nonirrigated grain corn production in Indiana were followed. Corn hybrid P0574AMXT was planted in 30-inch row spacing at a rate of 34,000 seeds/acre on May 22. All foliar fungicide applications were applied at 15 gal/acre and 40 psi using a Lee self-propelled sprayer equipped with a 10 feet boom, fitted with six TJ-VS 8002 nozzles spaced 20 inches apart. Fungicides were applied on July 15 at V12 growth stage and on July 26 at silk (R1) growth stage. Disease ratings were assessed on September 2 at dent (R5) growth stage. Gray leaf spot (GLS) and tar spot disease severity were visually assessed as a percentage (0–100%) of symptomatic leaf area on ear leaf, with five plants assessed per plot and ratings averaged before analysis. The two center rows of each plot were harvested on October 23, and yields were adjusted to 15.5% moisture. All data were analyzed in SAS 9.4 (SAS Institute, Cary, NC). A generalized linear mixed model analysis of variance was performed using PROC GLIMMIX. Values are least squares means, and values with different letters are significantly different based on least squares means test ($\alpha=0.05$).

In 2021, weather conditions were not favorable for disease. GLS was the most prominent disease in the trial and reached low severity. All fungicide programs significantly reduced GLS severity on the ear leaf compared to the nontreated controls on September 2 (Table 1). There was no significant difference between treatments for tar spot stroma severity, harvest moisture, test weight, and yield of corn.

TABLE 1. *Effect of Fungicide on Foliar Diseases Severity and Yield of Corn*

TREATMENT, RATE/ACRE, AND TIMING[z]	GLS[y] %	TAR SPOT[y] %	HARVEST MOISTURE %	TEST WEIGHT LB/BU	YIELD[x] BU/ACRE
Nontreated control 1	2.7 a	0.00	16.6	55.4	155.0
Headline AMP 1.68 SC 10.0 fl oz at V12	1.1 b-e	0.00	16.4	56.0	168.3
Veltyma 3.34 S 7.0 fl oz at V12	0.5 de	0.01	16.5	56.1	157.0
Trivapro 2.21 SE 13.7 fl oz at V12	0.6 de	0.00	16.4	55.9	170.8
Delaro Complete 458 SC 8.0 fl oz at V12	1.7 bc	0.02	16.6	55.4	175.9
Lucento 4.17 SC 5.0 fl oz at V12	0.8 cde	0.00	16.1	55.5	186.1
Nontreated control 2	1.9 ab	0.00	16.1	56.0	160.8
Headline AMP 1.68 SC 10.0 fl oz at R1	1.1 b-e	0.00	17.1	55.2	156.9
Veltyma 3.34 S 7.0 fl oz at R1	0.4 e	0.00	16.3	57.3	172.6
Trivapro 2.21 SE 13.7.0 fl oz at R1	1.4 bcd	0.00	16.4	55.6	163.2
Delaro Complete 458 SC 8.0 fl oz at R1	1.0 b-e	0.00	16.3	55.8	161.3
Lucento 4.17 SC 5.0 fl oz at R1	0.8 cde	0.00	16.9	55.6	163.7
P-value[w]	0.0014	0.4671	0.6977	0.1306	0.8274

[z] Fungicide treatments applied on July 15 at V12 growth stage and on July 26 at silk (R1) growth stage. All foliar treatments contained a nonionic surfactant (Preference) at a rate of 0.25% v/v.

[y] Foliar disease severity was visually assessed as a percentage (0–100%) of symptomatic leaf area on ear leaf, with five plants assessed per plot and ratings averaged before analysis on September 2. GLS = gray leaf spot.

[x] Yields were adjusted to 15.5% moisture after harvesting on October 23.

[w] All data were analyzed in SAS 9.4 (SAS Institute, Cary, NC). A generalized linear mixed model analysis of variance was performed using PROC GLIMMIX. Values are least squares means, and values with different letters are significantly different based on least squares means test (α=0.05).

EVALUATION OF FUNGICIDES FOR FOLIAR DISEASES IN CORN IN CENTRAL INDIANA, 2021 (COR21-12.ACRE)

S. Shim and D. E. P. Telenko, Department of Botany and Plant Pathology, Purdue University West Lafayette, IN 47907-2054

CORN (ZEA MAYS P0574AMXT)

Gray leaf spot, *Cercospora zeae-maydis*
Tar spot, *Phyllachora maydis*

A trial was established at the Purdue Agronomy Center for Research and Education (ACRE) in Tippecanoe County, Indiana. The experiment was a randomized complete block design with four replications. Plots were 10 feet wide and 30 feet long and consisted of four rows, and the two center rows were used for evaluation. The previous crop was corn. Standard practices for nonirrigated grain corn production in Indiana were followed. Corn hybrid P0574AMXT was planted in 30-inch row spacing at a rate of 34,000 seeds/acre on May 22. Foliar applications were made at silk (R1) growth stage on July 26. All foliar fungicide applications were applied at 15 gal/acre and 40 psi using a Lee self-propelled sprayer equipped with a 10-foot boom, fitted with six TJ-VS 8002 nozzles spaced 20-inches apart. Disease ratings were assessed on September 2 at dent (R5) growth stage. Gray leaf spot (GLS) and tar spot disease severity was visually assessed as a percentage (0–100%) of symptomatic leaf area on ear leaf, with five plants assessed per plot a α nd ratings averaged before analysis. The two center rows of each plot were harvested on October 23 and yields were adjusted to 15.5% moisture. All data were analyzed in SAS 9.4 (SAS Institute, Cary, NC). A generalized linear mixed model analysis of variance was performed using PROC GLIMMIX. Values are least squares means, and values with different letters are significantly different based on least squares means test ($\alpha=0.05$).

In 2021, weather conditions were not favorable for disease. GLS and tar spot were present in the trial but only remained at low levels. All treatments reduced GLS over the nontreated control on September 2 (Table 2). There was no significant effect of treatment on tar spot over the nontreated control. There was no significant effect of treatment on harvest moisture, test weight and yield of corn.

TABLE 2. *Effect of Treatment on Foliar Disease Severity and Yield of Corn*

TREATMENT AND RATE/ACRE[z]	GLS[y] %	TAR SPOT[y] %	HARVEST MOISTURE %	TEST WEIGHT LB/BU	YIELD[x] BU/ACRE
Nontreated control	1.8 a	0.00	16.9	55.7	160.5
Delaro Complete 485 SC 8.0 fl oz	0.4 bc	0.01	16.6	55.8	177.8
Veltyma 3.34 S 7.0 fl oz	0.1 c	0.01	17.2	55.2	167.3
Trivapro 2.21 SE 13.7 fl oz	0.4 bc	0.01	17.0	55.7	168.6
Miravis Neo 2.5 SE 13.7 fl oz	0.3 bc	0.00	16.7	55.6	176.1
Brixen 15.0 fl oz	0.3 bc	0.00	16.5	55.8	167.4
Brixen 13.0 fl oz	0.5 bc	0.01	16.4	55.7	167.6
Brixen 10.0 fl oz	0.2 c	0.00	16.4	55.8	167.9
Zolera ODX 5.0 fl oz	0.2 c	0.01	16.8	55.4	167.7
Aproach Prima 2.34 SC 6.8 fl oz	0.6 b	0.01	16.3	55.9	168.9
Brixen 10.0 fl oz + Proline 480 SC 1 fl oz	0.1 c	0.00	16.3	55.5	167.4
P-value[w]	*0.0001*	*0.8725*	*0.3601*	*0.9341*	*0.9839*

[z] Foliar applications were made at silk (R1) growth stage on July 26. All treatments contained a nonionic surfactant (Preference) at a rate of 0.25% v/v.

[y] Foliar disease severity was visually assessed as a percentage (0–100%) of symptomatic leaf area on ear leaf, with five plants assessed per plot and ratings averaged before analysis on September 2. GLS = gray leaf spot.

[x] Yields were adjusted to 15.5% moisture after harvesting on October 23.

[w] All data were analyzed in SAS 9.4 (SAS Institute, Cary, NC). A generalized linear mixed model analysis of variance was performed using PROC GLIMMIX. Values are least squares means, and values with different letters are significantly different based on least squares means test (α=0.05).

EVALUATION OF IN-FURROW FUNGICIDES IN CORN IN CENTRAL INDIANA, 2021 (COR21-20.ACRE)

S. Shim and D. E. P. Telenko, Department of Botany and Plant Pathology, Purdue University West Lafayette, IN 47907-2054

CORN (*ZEA MAYS* P0574AMXT)

Gray leaf spot, *Cercospora zeae-maydis*
Tar spot, *Phyllachora maydis*

A trial was established at the Purdue Agronomy Center for Research and Education (ACRE) in Tippecanoe County, Indiana. The experiment was a randomized complete block design with four replications. Plots were 10 feet wide and 30 feet long and consisted of four rows, and the two center rows were used for evaluation. The previous crop was corn. Standard practices for nonirrigated grain corn production in Indiana were followed. Corn hybrid P0574AMXT was planted in 30-inch row spacing at a rate of 2 seeds/foot on May 14. In-furrow applications were applied at planting at 10 gal/acre. Foliar applications were made at silk (R1) growth stage on July 26. All foliar fungicide applications were applied at 15 gal/acre and 40 psi using a Lee self-propelled sprayer equipped with a 10-foot boom, fitted with six TJ-VS 8002 nozzles spaced 20 inches apart. Disease ratings were assessed on September 2 at dent (R5) growth stage. Gray leaf spot (GLS) and tar spot disease severity were visually assessed as a percentage (0–100%) of symptomatic leaf area on ear leaf, with five plants assessed per plot and ratings averaged before analysis. The two center rows of each plot were harvested on October 23, and yields were adjusted to 15.5% moisture. All data were analyzed in SAS 9.4 (SAS Institute, Cary, NC). A generalized linear mixed model analysis of variance was performed using PROC GLIMMIX. Values are least squares means, and values with different letters are significantly different based on least squares means test (α=0.05).

In 2021, weather conditions were not favorable for disease. GLS and tar spot were present in the trial but only remained at low levels. All treatments reduced GLS severity over the nontreated control except Tepera in-furrow (Table 3). Treatments that included a fungicide application at R1 resulted in the lowest amount of GLS verses Xyway in-furrow only application. There was no significant effect of treatment on tar spot stroma severity, harvest moisture, test weight, and yield of corn.

TABLE 3. *Effect of Treatment on Folia Disease Severity and Yield of Corn*

TREATMENT, RATE/ACRE, AND TIMING[z]	GLS[y] %	TAR SPOT[y] %	HARVEST MOISTURE %	TEST WEIGHT LB/BU	YIELD[w] BU/ACRE
Nontreated control	4.0 a	0.0	16.0	54.4	146.9
Xyway LFR 15.2 fl oz in-furrow	2.6 bc	0.1	15.5	54.7	164.9
Xyway LFR 10.5 fl oz in-furrow fb Topguard EQ 4.29 SC 5.0 fl oz at R1	1.6 cd	0.0	15.6	55.3	165.1
Topguard EQ 4.29 SC 5.0 fl oz at R1	1.4 d	0.0	15.6	54.5	153.0
Veltym 3.34 SC 7.0 fl oz at R1	0.7 d	0.0	15.6	54.5	150.2
Tepera Plus HD 5.4 fl oz in-furrow	3.5 ab	0.0	15.9	55.1	153.3
P-value[w]	*0.0001*	*0.2431*	*0.6271*	*0.0790*	*0.7120*

[z] In-furrow applications were applied at planting on May 14. Foliar applications were made at silk (R1) growth stage on July 26. fb = followed by.

[y] Foliar disease severity was visually assessed as a percentage (0–100%) of symptomatic leaf area on ear leaf, with five plants assessed per plot and ratings averaged before analysis on September 2. GLS = gray leaf spot.

[x] Yields were adjusted to 15.5% moisture after harvesting on October 23.

[w] All data were analyzed in SAS 9.4 (SAS Institute, Cary, NC). A generalized linear mixed model analysis of variance was performed using PROC GLIMMIX. Values are least squares means, and values with different letters are significantly different based on least squares means test (α=0.05).

XYWAY EFFICACY FOR FOLIAR DISEASE IN CORN IN CENTRAL INDIANA, 2021 (COR21-24.ACRE)

S. Shim and D. E. P. Telenko, Department of Botany and Plant Pathology, Purdue University West Lafayette, IN 47907-2054

CORN (ZEA MAYS P0574AMXT)

Gray leaf spot, *Cercospora zeae-maydis*
Tar spot, *Phyllachora maydis*

A trial was established at the Purdue Agronomy Center for Research and Education (ACRE) in Tippecanoe County, Indiana. The experiment was a randomized complete block design with four replications. Plots were 10 feet wide and 30 feet long and consisted of four rows, and the two center rows were used for evaluation. The previous crop was corn. Standard practices for nonirrigated grain corn production in Indiana were followed. Corn hybrid P0574AMXT was planted in 30-inch row spacing at a rate of 2 seeds/foot on May 14. Xyway applications were applied at planting. Foliar applications were made at silk (R1) growth stage on July 26. All foliar fungicide applications were applied at 15 gal/acre and 40 psi using a Lee self-propelled sprayer equipped with a 10-foot boom, fitted with six TJ-VS 8002 nozzles spaced 20 inches apart. Disease ratings were assessed on September 2 at dent (R5) growth stage. Gray leaf spot (GLS) and tar spot disease severity were visually assessed as a percentage (0–100%) of symptomatic leaf area on ear leaf, with five plants assessed per plot and ratings averaged before analysis. The two center rows of each plot were harvested on October 23, and yields were adjusted to 15.5% moisture. All data were analyzed in SAS 9.4 (SAS Institute, Cary, NC). A generalized linear mixed model analysis of variance was performed using PROC GLIMMIX. Values are least squares means, and values with different letters are significantly different based on least squares means test (α=0.05).

In 2021, weather conditions were not favorable for disease. GLS and tar spot were present in the trial but only remained at low levels. All treatments reduced GLS over the nontreated control on September 2 (Table 4). There was no significant effect of treatment on tar spot stroma severity on September 2. Treatments that included Xyway in-furrow, dribbled at 3 and 7 gal/acre, and Delaro had reduced test weight over the nontreated control. There was no significant effect of treatment on canopy greenness, moisture, and yield of corn.

TABLE 4. *Effect of Treatment on Foliar Disease Severity and Yield of Corn*

TREATMENT, RATE/ACRE, AND TIMING[z]	GLS[y] %	TAR SPOT[y] %	CANOPY GREEN[x] %	HARVEST MOISTURE %	TEST WEIGHT LB/BU	YIELD[w] BU/ACRE
Nontreated control	3.5 a	0.00	51.3	16.0	55.5 a	141.3
Xyway LFR 15.2 fl oz in-furrow 10 GPA	1.9 b	0.00	61.3	16.3	54.6 c	133.9
Xyway LFR 15.2 fl oz 2x2 10 GPA	2.1 b	0.00	57.5	16.0	54.9 abc	128.6
Xyway LFR 15.2 fl oz Y-drop 2 stream on tee jet	2.2 b	0.00	60.0	16.3	55.4 ab	133.8
Xyway LFR 15.2 fl oz dribble 3 GPA single stream nozzle	2.0 b	0.00	53.8	15.9	54.4 c	130.4
Xyway LFR 15.2 fl oz dribble 5 GPA single stream nozzle	1.6 b	0.00	57.5	15.8	55.6 a	129.7
Xyway LFR 15.2 fl oz dribble 7 GPA single stream nozzle	1.9 b	0.00	58.8	16.0	54.7 bc	143.4
Xyway LFR 15.2 fl oz dribble 5 GPA 2 off-row single stream	2.1 b	0.01	55.0	15.8	55.4 ab	129.8
Delaro Complete 458 SC 8.0 fl oz at R1	1.3 b	0.00	58.8	16.2	54.8 bc	127.3
P-value[w]	*0.0264*	*0.4613*	*0.2159*	*0.7071*	*0.0124*	*0.9410*

[z] Xyway applications were applied at planting on May 14, and foliar applications were made at R1 (silk) stage on July 26. GPA = gallons per acre.

[y] Foliar disease severity was visually assessed as a percentage (0–100%) of symptomatic leaf area on ear leaf, with five plants assessed per plot and ratings averaged before analysis on September 2. GLS = gray leaf spot.

[x] Canopy greenness was visually assessed as a percentage (0–100%) of canopy green on September 2.

[x] Yields were adjusted to 15.5% moisture after harvesting on October 23. [w] All data were analyzed in SAS 9.4 (SAS Institute, Cary, NC). A generalized linear mixed model analysis of variance was performed using PROC GLIMMIX. Values are least squares means, and values with different letters are significantly different based on least squares means test (α=0.05).

XYWAY EFFICACY FOR STALK ROT DISEASES IN CORN IN CENTRAL INDIANA, 2021 (COR21-25.ACRE)

S. Shim and D. E. P. Telenko, Department of Botany and Plant Pathology, Purdue University West Lafayette, IN 47907-2054

CORN (*ZEA MAYS* W2585SSRIB)

Gray leaf spot, *Cercospora zeae-maydis*

Stalk rot, *Stenocarpella maydis, Colletotrichum graminicola, Fusarium graminearum, Nigrospora oryzae, Fusarium sp.*

A trial was established at the Purdue Agronomy Center for Research and Education (ACRE) in Tippecanoe County, Indiana. The experiment was a randomized complete block design with four replications. Plots were 10 feet wide and 30 feet long and consisted of four rows, and the two center rows were used for evaluation. The previous crop was corn. Standard practices for nonirrigated grain corn production in Indiana were followed. Corn hybrid W2585SSRIB was planted in 30-inch row spacing at a rate of 2 seeds/foot on May 14. Xyway applications were applied in-furrow at 10 gal/acre at planting. Foliar applications were made at silk (R1) growth stage on July 26. All foliar fungicide applications were applied at 15 gal/acre and 40 psi using a Lee self-propelled sprayer equipped with a 10-foot boom, fitted with six TJ-VS 8002 nozzles spaced 20 inches apart. Disease ratings were assessed on September 2 at dent (R5) growth stage. Gray leaf spot (GLS) disease severity was visually assessed as a percentage (0–100%) of symptomatic leaf area on ear leaf, with five plants assessed per plot and ratings averaged before analysis. The two center rows of each plot were harvested on October 23, and yields were adjusted to 15.5% moisture. All data were analyzed in SAS 9.4 (SAS Institute, Cary, NC). A generalized linear mixed model analysis of variance was performed using PROC GLIMMIX. Values are least squares means, and values with different letters are significantly different based on least squares means test (α=0.05).

In 2021, weather conditions were not favorable for disease. GLS was present in the trial but only remained at low levels. Stalk disease was evaluated, and stalk rot pathogens identified included *Stenocarpella maydis, Colletotrichum graminicola, Fusarium graminearum, Nigrospora oryzae,* and *Fusarium spp.* All treatments reduced GLS over the nontreated control on September 2 except Xyway at 7.6 oz in-furrow (Table 5). Xyway 15.2 fl oz in-furrow and Xyway 10.5 fl oz in-furrow followed by Topguard 5 fl oz at R1 increased % canopy green over the nontreated control on September 2. Treatments that included Xyway in-furrow reduced stalk disease over the nontreated control. There was no significant effect of treatment on harvest moisture, test weight, and yield of corn.

TABLE 5. *Effect of Treatment on Foliar Disease Severity and Yield of Corn*

TREATMENT, RATE/ACRE, AND TIMING[z]	GLS[y] %	CANOPY GREEN[x] %	STALK[w] %	HARVEST MOISTURE %	TEST WEIGHT LB/BU	YIELD[w] BU/ACRE
Nontreated control	2.7 a	50.0 b	3.7 a	16.1	55.4	167.1
Xyway LFR 15.2 fl oz in-furrow	1.4 b	61.3 a	2.6 bc	16.6	54.9	157.5
Xyway LFR 10.5 fl oz in-furrow fb Topguard EQ 5.0 fl oz at R1	1.7 b	61.7 a	2.4 c	16.8	55.1	166.3
Topguard EQ 5.0 fl oz at R1	1.4 b	56.7 ab	3.1 ab	16.1	55.3	167.7
Xyway LFR 7.6 fl oz in-furrow	1.9 ab	56.7 ab	2.4 bc	16.2	55.0	160.3
P-value[u]	*0.0201*	*0.0459*	*0.0039*	*0.3098*	*0.8220*	*0.8762*

[z] Xyway applications were applied in-furrow, 2x2, and dribbled by hand at 10 gal/acre at planting on May 15, and foliar applications were made at silk (R1) growth stage on July 26 and contained a nonionic surfactant (Preference) at a rate of 0.25% v/v. fb = followed by.

[y] Gray leaf spot disease severity was visually assessed as a percentage (0–100%) of symptomatic leaf area on ear leaf, with five plants assessed per plot, and ratings averaged before analysis on September 2. GLS = gray leaf spot.

[x] Canopy greenness was visually assessed as a percentage (0–100%) of canopy green on September 2.

[w] Stalk disease was rated on October 11. A scale of 0–5 (Hines, University of Illinois) was used in which 0 = no visible discoloration of the internal below ear stalk nodes or pith, 1 = internal discoloration at the stalk nodes below the ear, 2 = internal discoloration at the stalk nodes and in the pith below the ear, 3 = pith separation occurring below the ear, 4 = complete discoloration and decay of the pith between at least two nodes below the ear but stalk still standing, and 5 = stalk lodged below the ear due to stalk rot.

[v] Yields were adjusted to 15.5% moisture after harvesting on October 23.

[u] All data were analyzed in SAS 9.4 (SAS Institute, Cary, NC). A generalized linear mixed model analysis of variance was performed using PROC GLIMMIX. Values are least squares means, and values with different letters are significantly different based on least squares means test (α=0.05).

FUNGICIDE COMPARISON FOR FOLIAR DISEASES IN CORN IN CENTRAL INDIANA, 2021 (COR21-26.ACRE)

S. Shim and D. E. P. Telenko, Department of Botany and Plant Pathology, Purdue University West Lafayette, IN 47907-2054

CORN (*ZEA MAYS* P0574AMXT)

Gray leaf spot, *Cercospora zeae-maydis*
Tar spot, *Phyllachora maydis*

A trial was established at the Purdue Agronomy Center for Research and Education (ACRE) in Tippecanoe County, Indiana. The experiment was a randomized complete block design with four replications. Plots were 10 feet wide and 30 feet long and consisted of four rows, and the two center rows were used for evaluation. The previous crop was corn. Standard practices for nonirrigated grain corn production in Indiana were followed. Corn hybrid P0574AMT was planted in 30-inch row spacing at a rate of 2 seeds/foot on May 14. Xyway applications were applied in-furrow at 10 gal/acre at planting. Foliar applications were made at silk (R1) growth stage on July 26. All foliar fungicide applications were applied at 15 gal/acre and 40 psi using a Lee self-propelled sprayer equipped with a 10-foot boom, fitted with six TJ-VS 8002 nozzles spaced 20 inches apart. Disease ratings were assessed on September 2 at dent (R2) growth stage. Gray leaf spot (GLS) and tar spot disease severity were visually assessed as a percentage (0–100%) of symptomatic leaf area on ear leaf, with five plants assessed per plot and ratings averaged before analysis. The two center rows of each plot were harvested on October 23, and yields were adjusted to 15.5% moisture. All data were analyzed in SAS 9.4 (SAS Institute, Cary, NC). A generalized linear mixed model analysis of variance was performed using PROC GLIMMIX. Values are least squares means, and values with different letters are significantly different based on least squares means test (α=0.05).

In 2021, weather conditions were not favorable for disease. GLS and tar spot were present in the trial but only remained at low levels. All treatments reduced GLS over the nontreated control on September 2 (Table 6). There was no significant effect of tar spot stroma severity, percent canopy green, harvest moisture, test weight, and yield of corn.

TABLE 6. *Effect of Treatment on Foliar Disease Severity and Yield of Corn*

TREATMENT, RATE/ACRE, AND TIMING[z]	GLS[y] %	TAR SPOT[y] %	CANOPY GREEN[x] %	HARVEST MOISTURE %	TEST WEIGHT LB/BU	YIELD[w] BU/ACRE
Nontreated control	4.2 a	0.01	53.8	15.6	54.7	156.1
Topguard EQ 4.29 5.0 fl oz at R1	2.2 bc	0.01	61.9	15.7	66.2	159.4
Lucento 4.17 SC 5.0 fl oz at R1	2.5 b	0.00	60.0	15.7	55.3	157.4
Veltyma 3.34 S 7.0 fl oz at R1	1.4 c	0.01	65.0	16.2	55.6	158.3
Delaro Complete 458 SC 8.0 fl oz at R1	2.3 bc	0.00	60.0	15.5	55.6	154.4
Miravis Neo 2.4 SE 13.7 fl oz at R1	1.4 c	0.00	58.8	15.6	55.3	168.7
Xyway LFR 15.2 fl oz in-furrow	2.9 b	0.06	58.8	15.4	55.3	164.3
P-value[v]	0.0009	0.5063	0.3119	0.1533	0.2382	0.5542

[z] Xyway applications were applied in-furrow at 10 gal/acre at planting on May 15, and foliar applications were made at silk (R1) growth stage on July 26. All foliar treatments contained a nonionic surfactant (Preference) at a rate of 0.25% v/v.

[y] Foliar disease severity was visually assessed as a percentage (0–100%) of symptomatic leaf area on ear leaf, with five plants assessed per plot and ratings averaged before analysis on September 2. GLS = gray leaf spot.

[x] Canopy greenness was visually assessed as a percentage (0–100%) of canopy green on September 2.

[w] Yields were adjusted to 15.5% moisture after harvesting on October 23.

[v] All data were analyzed in SAS 9.4 (SAS Institute, Cary, NC). A generalized linear mixed model analysis of variance was performed using PROC GLIMMIX. Values are least squares means, and values with different letters are significantly different based on least squares means test (α=0.05).

FUNGICIDE COMPARISON FOR FOLIAR DISEASES IN CORN IN CENTRAL INDIANA, 2021 (COR21-28.ACRE)

S. Shim and D. E. P. Telenko, Department of Botany and Plant Pathology, Purdue University West Lafayette, IN 47907-2054

CORN (*ZEA MAYS* P0574AMXT)

Gray leaf spot, *Cercospora zeae-maydis*
Tar spot, *Phyllachora maydis*

A trial was established at the Purdue Agronomy Center for Research and Education (ACRE) in Tippecanoe County, Indiana. The experiment was a randomized complete block design with four replications. Plots were 10 feet wide and 30 feet long and consisted of four rows, and the two center rows were used for evaluation. The previous crop was corn. Standard practices for nonirrigated grain corn production in Indiana were followed. Corn hybrid P0574AMXT was planted in 30-inch row spacing at a rate of 34,000 seeds/acre on May 22. Foliar applications were made at V5, V12, and silk (R1) growth stages on June 24, July 15, and July 26, respectively. All foliar fungicide applications were applied at 15 gal/acre and 40 psi using either a CO_2 backpack sprayer or a Lee self-propelled sprayer equipped with a 10-foot boom, fitted with six TJ-VS 8002 nozzles spaced 20 inches apart. Disease ratings were assessed on September 2 at dent (R5) growth stage. Gray leaf spot (GLS) and tar spot disease severity were visually assessed as a percentage (0–100%) of symptomatic leaf area on ear leaf with five plants assessed per plot and ratings averaged before analysis. The two center rows of each plot were harvested on October 23, and yields were adjusted to 15.5% moisture. All data were analyzed in SAS 9.4 (SAS Institute, Cary, NC). A generalized linear mixed model analysis of variance was performed using PROC GLIMMIX. Values are least squares means, and values with different letters are significantly different based on least squares means test (α=0.05).

In 2021, weather conditions were not favorable for disease. GLS and tar spot were present in the trial but only remained at low levels. All treatments reduced GLS over the nontreated control on September 2 except Delaro Complete 4.0 fl oz at V5 (Table 7). There was no significant effect of treatment on tar spot stroma severity on September 2. There was no significant effect of treatment on harvest moisture, test weight, and yield of corn.

TABLE 7. *Effect of Treatment on Foliar Disease Severity and Yield of Corn*

TREATMENT, RATE/ACRE, AND TIMING[z]	GLS[y] %	TAR SPOT[y] %	HARVEST MOISTURE %	TEST WEIGHT LB/BU	YIELD[x] BU/ACRE
Nontreated control	0.7 a	0.02	17.0	55.0	167.9
Delaro Complete 458 SC 4.0 fl oz at V5	0.7 a	0.00	16.6	55.4	170.4
Delaro Complete 458 SC 8.0 fl oz at V12	0.1 b	0.03	16.1	55.7	177.8
Delaro Complete 458 SC 12.0 fl oz at V12	0.1 b	0.02	16.7	55.6	172.5
Veltyma 3.34 S 7.0 fl oz at V12	0.2 b	0.01	16.7	56.0	174.1
Miravis Neo 2.4 SE 13.0 fl oz at V12	0.2 b	0.01	17.2	55.4	181.5
Delaro Complete 458 SC 12.0 fl oz at R1	0.2 b	0.01	17.1	54.8	177.9
Veltyma 3.34 S 7.0 fl oz at R1	0.0 b	0.01	16.9	55.2	171.7
Miravis Neo 2.4 SE 13.0 fl oz at R1	0.1 b	0.00	16.5	55.2	173.2
P-value[w]	0.0037	0.1197	0.6691	0.4470	0.9099

[z] Foliar applications were made on June 24, July 15, and July 26 at V5, V12, and silk (R1) growth stages, respectively. All treatments applied at R1 contained a nonionic surfactant (Preference) at a rate of 0.25% v/v.

[y] Foliar disease severity was visually assessed as a percentage (0–100%) of symptomatic leaf area on ear leaf, with five plants assessed per plot and ratings averaged before analysis on September 2. GLS = gray leaf spot.

[x] Yields were adjusted to 15.5% moisture after harvesting on October 23.

[w] All data were analyzed in SAS 9.4 (SAS Institute, Cary, NC). A generalized linear mixed model analysis of variance was performed using PROC GLIMMIX. Values are least squares means, and values with different letters are significantly different based on least squares means test (α=0.05).

FUNGICIDE COMPARISON AT V5 FOR FOLIAR DISEASES IN CORN IN CENTRAL INDIANA, 2021 (COR21-32.ACRE)

S. Shim and D. E. P. Telenko, Department of Botany and Plant Pathology, Purdue University West Lafayette, IN 47907-2054

CORN (ZEA MAYS P0574AMXT)

Gray leaf spot, *Cercospora zeae-maydis*
Tar spot, *Phyllachora maydis*

A trial was established at the Purdue Agronomy Center for Research and Education (ACRE) in Tippecanoe County, Indiana. The experiment was a randomized complete block design with four replications. Plots were 10 feet wide and 30 feet long and consisted of four rows, and the two center rows were used for evaluation. The previous crop was corn. Standard practices for nonirrigated grain corn production in Indiana were followed. Corn hybrid P0574AMXT was planted in 30-inch row spacing at a rate of 34,000 seeds/acre on May 22. Foliar applications were made at V5 growth stage on June 24. All foliar fungicide applications were applied at 15 gal/acre and 40 psi using a CO_2 backpack sprayer equipped with a 10-foot boom, fitted with six TJ-VS 8002 nozzles spaced 20 inches apart. Disease ratings were assessed on September 2 at dent (R5) growth stage. Gray leaf spot (GLS) and tar spot disease severities were visually assessed as a percentage (0–100%) of symptomatic leaf area on ear leaf, with five plants assessed per plot and ratings averaged before analysis The two center rows of each plot were harvested on October 23, and yields were adjusted to 15.5% moisture All data were analyzed in SAS 9.4 (SAS Institute, Cary, NC). A generalized linear mixed model analysis of variance was performed using PROC GLIMMIX. Values are least squares means, and values with different letters are significantly different based on least squares means test (α=0.05).

In 2021, weather conditions were not favorable for disease. GLS and tar spot were present in the trial but only remained at low levels. There was no significant effect of treatment on GLS and tar spot stroma severity over the nontreated control on September 2 (Table 8). There was no significant effect of treatment on harvest moisture, test weight, and yield of corn.

TABLE 8. *Effect of Treatment on Foliar Disease Severity and Yield of Corn*

TREATMENT, RATE/ACRE, AND TIMING[z]	GLS[y] %	TAR SPOT[y] %	HARVEST MOISTURE %	TEST WEIGHT LB/BU	YIELD[x] BU/ACRE
Nontreated control	1.8	0.00	16.6	55.3	145.5
Affiance 1.5 SC 10.0 fl oz at V5	1.1	0.00	16.6	55.5	144.1
Domark 230 ME 5.0 fl oz at V5	1.0	0.01	16.5	55.6	141.5
Revytek 3.33 LC SC 8.0 fl oz at V5	1.1	0.01	16.1	55.7	143.8
Miravis Neo 2.4 SE 13.7 fl oz at V5	1.3	0.00	15.9	55.9	140.1
Trivapro 2.21 SE 13.7 fl oz at V5	1.7	0.00	16.1	55.9	144.7
Veltyma 3.34 S 7.0 fl oz at V5	1.4	0.00	16.3	56.0	136.5
Delaro Complete 458 SC 8.0 fl oz at V5	1.0	0.00	16.2	56.2	139.6
P-value[w]	*0.1539*	*0.5828*	*0.5482*	*0.5021*	*0.9158*

[z] Foliar applications were made at V5 growth stage on June 24.

[y] Foliar disease severity was visually assessed as a percentage (0–100%) of symptomatic leaf area on ear leaf, with five plants assessed per plot and ratings averaged before analysis on September 2. GLS = gray leaf spot.

[x] Yields were adjusted to 15.5% moisture after harvesting on October 23.

[w] All data were analyzed in SAS 9.4 (SAS Institute, Cary, NC). A generalized linear mixed model analysis of variance was performed using PROC GLIMMIX. Values are least squares means, and values with different letters are significantly different based on least squares means test (α=0.05).

EVALUATION OF OROAGRI PRODUCTS IN CORN IN CENTRAL INDIANA, 2021 (COR21-34.ACRE)

S. Shim and D. E. P. Telenko, Department of Botany and Plant Pathology, Purdue University West Lafayette, IN 47907-2054

CORN (*ZEA MAYS* P0574AMXT)

Gray leaf spot, *Cercospora zeae-maydis*

A trial was established at the Purdue Agronomy Center for Research and Education (ACRE) in Tippecanoe County, Indiana. The experiment was a randomized complete block design with four replications. Plots were 10 feet wide and 30 feet long and consisted of four rows, and the two center rows were used for evaluation. The previous crop was corn. Standard practices for nonirrigated grain corn production in Indiana were followed. Corn hybrid P0574AMXT was planted in 30-inch row spacing at a rate of 34,000 seeds/acre on May 22. Foliar applications were made at silk (R1) growth stage on July 26. All foliar fungicide applications were applied at 15 gal/acre and 40 psi using a Lee self-propelled sprayer equipped with a 10-foot boom, fitted with six TJ-VS 8002 nozzles spaced 20 inches apart. Disease ratings were assessed on September 2 at dent (R5) growth stage. Gray leaf spot (GLS) was visually assessed as a percentage (0–100%) of symptomatic leaf area on ear leaf, with five plants assessed per plot and ratings averaged before analysis on September 2. The two center rows of each plot were harvested on October 23, and yields were adjusted to 15.5% moisture. All data were analyzed in SAS 9.4 (SAS Institute, Cary, NC). A generalized linear mixed model analysis of variance was performed using PROC GLIMMIX. Values are least squares means, and values with different letters are significantly different based on least squares means test (α=0.05).

In 2021, weather conditions were not favorable for disease. GLS was present in the trial but only remained at low levels. All treatments reduced GLS as compared to the nontreated control (Table 9). There was no significant effect of treatment on harvest moisture, test weight, and yield of corn.

TABLE 9. *Effect of Treatment on Foliar Disease Severity and Yield of Corn*

TREATMENT AND RATE/ACRE[z]	GLS[y] %	HARVEST MOISTURE %	TEST WEIGHT LB/BU	YIELD[x] BU/ACRE
Nontreated control	1.18 a	16.4	55.4	166.7
Veltyma 3.34 S 7.0 fl oz	0.08 b	17.0	55.2	170.3
Veltyma 3.34 S 7.0 fl oz + OR-099-E 0.25% v/v	0.03 b	17.0	55.0	160.7
Veltyma 3.34 S 7.0 fl oz + OR-295-A 0.25% v/v	0.05 b	17.1	54.7	165.8
Veltyma 3.34 S 7.0 fl oz + OR-025-F 1.0 pt	0.01 b	17.1	55.1	162.8
Veltyma 3.34 S 7.0 fl oz + OR-009-A 0.4 % v/v	0.01 b	16.8	54.9	173.5
P-value[w]	0.0001	0.5737	0.6161	0.8333

[z] Foliar applications were made at V_5 growth stage on June 24.

[y] Gray leaf spot severity was visually assessed as a percentage (0–100%) of symptomatic leaf area on ear leaf, with five plants assessed per plot and ratings averaged before analysis on September 2. GLS = gray leaf spot.

[x] Yields were adjusted to 15.5% moisture after harvesting on October 23.

[w] All data were analyzed in SAS 9.4 (SAS Institute, Cary, NC). A generalized linear mixed model analysis of variance was performed using PROC GLIMMIX. Values are least squares means, and values with different letters are significantly different based on least squares means test (α=0.05).

IN-FURROW FUNGICIDE EVALUATION IN CORN IN CENTRAL INDIANA, 2021 (COR21-32.ACRE)

S. Shim and D. E. P. Telenko, Department of Botany and Plant Pathology, Purdue University West Lafayette, IN 47907-2054

CORN (*ZEA MAYS* P0574AMXT)

Gray leaf spot, *Cercospora zeae-maydis*
Tar spot, *Phyllachora maydis*

A trial was established at the Purdue Agronomy Center for Research and Education (ACRE) in Tippecanoe County, Indiana. The experiment was a randomized complete block design with four replications. Plots were 10 feet wide and 30 feet long and consisted of four rows, and the two center rows were used for evaluation. The previous crop was corn. Standard practices for nonirrigated grain corn production in Indiana were followed. Corn hybrid P0574AMXT was planted in 30-inch row spacing at a rate of 2 seeds/foot on May 14. In-furrow and 2x2 applications were applied at planting at 10 gal/acre. Disease ratings were assessed on September 2 at dent (R_5) growth stage. Gray leaf spot (GLS) and tar spot disease severities were visually assessed as a percentage (0–100%) of symptomatic leaf area on ear leaf, with five plants assessed per plot and ratings averaged before analysis. The two center rows of each plot were harvested on October 23, and yields were adjusted to 15.5% moisture. All data were analyzed in SAS 9.4 (SAS Institute, Cary, NC). A generalized linear mixed model analysis of variance was performed using PROC GLIMMIX. Values are least squares means, and values with different letters are significantly different based on least squares means test (α=0.05).

In 2021, weather conditions were not favorable for disease. GLS and tar spot were present in the trial but only remained at low levels. There was no significant effect of treatment on GLS and tar spot severity on September 2 (Table 10). There was no significant effect of treatment on percent canopy green, harvest moisture, test weight, and yield of corn.

TABLE 10. *Effect of Treatment on Foliar Disease Severity and Yield of Corn*

TREATMENT AND RATE/ACRE[z]	GLS[y] %	TAR SPOT[y] %	CANOPY GREEN[x] %	HARVEST MOISTURE %	TEST WEIGHT LB/BU	YIELD[w] BU/ACRE
Nontreated control	3.8	0.01	48.8	15.6	54.7	152.8
Double Nickel LC 8.0 oz in-furrow	2.2	0.00	41.7	15.0	54.0	151.6
Double Nickel LC 8.0 oz 2x2	2.7	0.00	47.5	15.2	54.8	150.8
Double Nickel LC 16.0 oz in-furrow	2.9	0.00	53.8	15.2	55.0	159.7
Double Nickel LC 16.0 oz 2x2	2.5	0.00	57.5	15.3	55.0	151.3
P-value[v]	0.4893	0.4860	0.2825	0.4781	0.8045	0.8408

[z] In-furrow and 2x2 applications were applied at planting on May 14.

[y] Foliar disease severity was visually assessed as a percentage (0–100%) of symptomatic leaf area on ear leaf, with five plants assessed per plot and ratings averaged before analysis on September 2. GLS = gray leaf spot.

[x] Canopy greenness was visually assessed as a percentage (0–100%) of canopy green on September 2.

[w] Yields were adjusted to 15.5% moisture after harvesting on October 23.

[v] All data were analyzed in SAS 9.4 (SAS Institute, Cary, NC). A generalized linear mixed model analysis of variance was performed using PROC GLIMMIX. Values are least squares means, and values with different letters are significantly different based on least squares means test (α=0.05).

FUNGICIDE COMPARISON FOR FOLIAR SOYBEAN DISEASES IN CENTRAL INDIANA, 2021 (SOY21-01.ACRE)

S. Shim and D. E. P. Telenko, Department of Botany and Plant Pathology, Purdue University West Lafayette, IN 47907-2054

SOYBEAN (*GLYCINE MAX* P34A79X)

Frogeye leaf spot, *Cercospora sojina*
Septoria brown spot, *Septoria glycines*

A trial was established at the Purdue Agronomy Center for Research and Education (ACRE) in Tippecanoe County, Indiana. The experiment was a randomized complete block design with four replications. Plots were 10 feet wide and 30 feet long and consisted of four rows, and the two center rows were used for evaluation. The previous crop was corn. Standard practices for soybean production in Indiana were followed. Soybean cultivar P35T15E was planted in 30-inch row spacing at a rate of 140,000 seeds/acre on May 22. Fungicide applications were applied on July 31 at beginning pod (R3) growth stage and were applied at 15 gal/acre at 40 psi using a CO_2 backpack sprayer equipped with a 10-foot boom, fitted with six TJ-VS 8002 nozzles spaced 20 inches apart. Disease ratings were assessed on September 8 at full seed (R6) growth stage. Frogeye leaf spot (FLS) and Septoria brown spot (SBS) were rated for disease severity by visually assessing the percentage of symptomatic leaf area in the upper and lower canopies, respectively. The two center rows of each plot were harvested on October 10, and yields were adjusted to 13% moisture. All data were analyzed in SAS 9.4 (SAS Institute, Cary, NC). A generalized linear mixed model analysis of variance was performed using PROC GLIMMIX. Values are least squares means, and values with different letters are significantly different based on least squares means test ($\alpha=0.05$).

In 2021, weather conditions were not favorable for soybean disease. FLS and SBS were present in the trial but only remained at low levels. All fungicides reduced SBS over the nontreated control on September 8 (Table 11). There was no significant effect of treatment on FLS severity, harvest moisture, test weight, and yield of soybean.

TABLE 11. *Effect of Treatment on Foliar Disease Severity and Yield of Soybean*

TREATMENT AND RATE/ACRE[z]	FLS[Y] %	SBS[Y] %	HARVEST MOISTURE %	TEST WEIGHT LB/BU	YIELD[x] BU/ACRE
Nontreated control	1.2	1.5 a	15.1	55.5	54.9
Preemptor 3.22 SC 5.0 fl oz	0.0	0.2 b	15.1	54.2	51.1
Topguard EQ 4.29 SC 5.0 fl oz	0.0	0.3 b	14.9	55.0	50.5
Quadris Top SBX 3.76 SC 7.0 fl oz	0.0	0.2 b	15.2	55.6	51.8
Lucento 4.17 SC 5.0 fl oz	0.0	0.4 b	15.4	54.6	55.5
Miravis Top 1.67 SC 13.7 fl oz	0.0	0.1 b	15.6	54.9	59.4
Priaxor Xemium SC 4.0 fl oz	0.0	0.2 b	15.4	54.3	52.7
Trivapro 2.21 SE 13.0 fl oz	0.0	0.1 b	15.2	54.4	57.8
Delaro 325 SC 8.0 fl oz	0.0	0.4 b	15.5	54.2	61.2
Headline 2.09 SC 10.0 fl oz	0.2	0.4 b	15.8	55.2	54.9
Veltyma 3.24 S 7.0 fl oz	0.0	0.2 b	15.3	54.4	57.4
Revytek 3.33 LC 8.0 fl oz	0.0	0.1 b	15.5	54.8	56.4
P-value[w]	0.2303	0.0001	0.6985	0.8217	0.2976

[z] Fungicide applications were made on July 13 at full seed (R3) growth stage and contained a nonionic surfactant (Preference) at a rate of 0.25% v/v.

[y] Foliar disease incidence was rated on a scale of 0–100% of plants within a plot with disease symptoms on September 8. FLS = frogeye leaf spot, SBS = Septoria brown spot.

[x] Yields were adjusted to 13% moisture after harvesting on October 10.

[w] All data were analyzed in SAS 9.4 (SAS Institute, Cary, NC). A generalized linear mixed model analysis of variance was performed using PROC GLIMMIX. Values are least squares means, and values with different letters are significantly different based on least squares means test (α=0.05).

EVALUATION OF SEED TREATMENTS AGAINST SDS AND SOYBEAN CYST NEMATODE ON SOYBEAN IN CENTRAL INDIANA, 2021 (SOY21-15.ACRE)

M. T. Brown, S. Shim, and D. E. P. Telenko, Department of Botany and Plant Pathology, Purdue University West Lafayette, IN 47907-2054

SOYBEAN (*GLYCINE MAX* P25T09E)

Sudden death syndrome, *Fusarium virguliforme*
Soybean cyst nematode, *Heterodera glycines*

A trial was established at the Purdue Agronomy Center for Research and Education (ACRE) in Tippecanoe County, Indiana. The experiment was a randomized complete block design with four replications. Plots were 10 feet wide and 30 feet long and consisted of four rows, and the two center rows were used for evaluation. The previous crop was soybean. Standard practices for soybean production in Indiana were followed. Soybean cultivar P25T09E was planted in 30-inch row spacing at a rate of 8 seeds/foot on May 22. Seed treatments were applied on seeds before planting; all treatments contained a base treatment except the nontreated control. Soybean cyst nematode (SCN) females were counted on June 21 at third/fourth trifoliate (V3/VC4) growth stages. White or tan females were extracted from the roots by washing over a #20 sieve nested over a #60 sieve and then counted using a dissecting microscope. SCN egg count was assessed on May 22 at planting and on September 27 at harvest. Five soybean roots were dug from each outside row, washed, and root rot was rated by visually assessing dark discoloration on roots on August 5 at full pod (R4) growth stage. The two center rows were harvested on September 27, and yields were adjusted to 13% moisture. All data were analyzed in SAS 9.4 (SAS Institute, Cary, NC). A generalized linear mixed model analysis of variance was performed using PROC GLIMMIX. Values are least squares means, and values with different letters are significantly different based on least squares means test (α=0.05).

In 2021, weather conditions were not favorable for soybean disease. No foliar symptoms were observed for soybean sudden death syndrome (SDS). No significant differences were observed between seed treatments and the nontreated control for SCN egg counts on May 22 and September 27 (Table 12). No significant differences between seed treatments and the nontreated control were found for SCN females on June 21 and for root rot on August 5. No significant differences between seed treatments and the nontreated control were found for yield of soybean.

TABLE 12. *Effect of Nematicide Seed Treatments against Soybean Cyst Nematode (SCN), Root Rot, and Yield of Soybean*

TREATMENT[z]	SCN FEMALES[y] JUN 21	SCN EGGS[x] MAY 22	SCN EGGS[x] SEP 27	ROOT ROT %[w] AUG 5	HARVEST MOISTURE %	TEST WEIGHT LB/BU	YIELD[v] BU/ACRE
Nontreated control	3.8	5,750	7,188	26.6	10.5	53.2	52.0
Base	1.8	4,625	5,063	28.8	10.8	54.4	55.1
Base + BioST 195.0 ml/100kg	5.5	4,500	4,250	29.9	10.0	54.0	53.5
Base + Aveo 2.0 ml/100000 seed	5.0	3,625	5,688	28.3	10.5	54.4	58.5
Base + Clariva PN 130.0 ml/100 kg	6.8	3,000	6,406	33.9	10.3	54.5	55.6
Base + ILeVO 0.15 mg/seed	3.8	4,563	4,031	24.8	10.5	55.0	52.6
Base + Trunemco 20.2 ml/100 kg	5.5	4,875	5,375	26.5	10.8	54.1	61.8
Base + Saltro 0.075 mg/seed	6.0	2,813	6,500	34.0	10.3	54.9	55.5
P-value[u]	0.3369	0.6265	0.8209	0.2469	0.3374	0.1121	0.0869

[z] Seed treatments applied before planting on May 22, all treatments contained a base treatment of Allegiance Fl at 4.0 g a/100 kg, Stamina at 7.5 g a/100 kg, Systiva XS Xemium Brand at 5.0 g a/100 kg, Poncho 600 at 0.11 mg a/seed, Flo Rite 1706 at 66.0 ml/100 kg, and Color Coat Red at 33.0 ml/100 kg except the nontreated control.

[y] SCN female was visually assessed as the number of white or tan females on June 21.

[x] SCN egg count was assessed on May 22 at planting and on September 27 at harvest for each treatment plot from soil samples.

[w] Root rot was visually assessed as a percentage (0–100%) of dark discoloration on roots on August 5.

[v] Yields were adjusted to 13% moisture after harvesting on September 27.

[u] All data were analyzed in SAS 9.4 (SAS Institute, Cary, NC). A generalized linear mixed model analysis of variance was performed using PROC GLIMMIX. Values are least squares means, and values with different letters are significantly different based on least squares means test (α=0.05).

EVALUATION OF THE EFFICACY OF SEED TREATMENTS IN SOYBEAN IN CENTRAL INDIANA, 2021 (SOY21-17.ACRE)

S. Shim and D. E. P. Telenko, Department of Botany and Plant Pathology, Purdue University West Lafayette, IN 47907-2054

SOYBEAN (*GLYCINE MAX* P28T14E AND P25A04X)

Sudden death syndrome, *Fusarium virguliforme*

A trial was established at the Purdue Agronomy Center for Research and Education (ACRE) in Tippecanoe County, Indiana. The experiment was a randomized complete block design with four replications. Plots were 10 feet wide and 30 feet long and consisted of four rows, and the two center rows were used for evaluation. The previous crop was corn. Standard practices for soybean production in Indiana were followed. Soybean cultivar P25A04X (resistant) and P28T14E (susceptible) were planted in 30-inch row spacing at a rate of 8 seeds/foot on May 15. Seed treatments were applied on seeds before planting. Ten roots per plot were sampled from border rows at full pod/beginning of seed (R4/R5) on August 9, gently washed, and rated for root rot severity on scale of 0–100%. Disease ratings were assessed on August 25 at maturity (R6). Sudden death syndrome (SDS) in each plot was rated for disease incidence (DI) and disease severity (DS). Disease incidence refers to the percentage of plants with disease symptoms, and disease severity (DS) was rated using a 1–9 scale where 1 refers to low disease pressure and 9 refers to premature death of the plant. SDS index was then calculated using the equation DX= (DI x DS)/9. The two center rows of each plot were harvested on October 11 and October 18—part of the harvest was delayed due to rain—and yields were adjusted to 13% moisture All data were analyzed in SAS 9.4 (SAS Institute, Cary, NC). A generalized linear mixed model analysis of variance was performed using PROC GLIMMIX. Values are least squares means, and values with different letters are significantly different based on least squares means test (α=0.05).

In 2021, weather conditions were not favorable for soybean disease. SDS was the most prominent diseases in the trial but only reached low severity. There was no significant difference between varieties and seed treatments for root rot on August 9. P25A04X nontreated and treated with Ilevo and Saltro had significantly lower levels of SDS incidence, severity, and index over nontreated P28T14E but was not significantly different from P28T14E treated with either Illevo or Saltro (Table 13). P25A04X has lower harvest moisture than P28T14E. There were no significant differences between cultivar and seed treatments for test weight and yield of soybean.

TABLE 13. *Effect of Seed Treatment on SDS, Root Rot, and Yield of Soybean*

CULTIVAR AND TREATMENT[z]	ROOT ROT %[x] AUG 9	SDS DI[y] AUG 25	SDS DS[y] AUG 25	SDS INDEX[y] AUG 25	HARVEST MOISTURE %	TEST WEIGHT LB/BU	YIELD[w] BU/ACRE
P25A04X, Nontreated control	13.5	0.3 b	0.3 b	0.0 b	13.0 c	53.8	63.9
P25A04X, ILeVO	18.9	0.3 b	0.3 b	0.0 b	13.1 bc	54.5	59.0
P25A04X, Saltro	10.8	0.5 b	0.5 ab	0.1 b	13.1 bc	55.0	59.1
P28T14E, Nontreated control	16.0	4.8 a	1.0 a	0.5 a	14.1 a	54.2	59.8
P28T14E, ILeVO	16.3	2.5 ab	1.0 a	0.3 ab	13.9 ab	53.9	56.4
P28T14E, Saltro	15.5	2.5 ab	1.0 a	0.3 ab	13.8 ab	53.5	65.5
P-value[v]	0.1423	0.0037	0.0179	0.0039	0.0342	0.3762	0.7465

[z] Seed treatments were preapplied to the seed of varieties P25A04X (resistant) and P28T14E (susceptible).

[y] Sudden death syndrome (SDS) in each plot was rated for disease incidence (DI) and disease severity (DS) on August 25. Disease incidence refers to the percentage of plants with disease symptoms, and disease severity (DS) was rated using a 1–9 scale where 1 refers to low disease pressure and 9 refers to premature death of the plant.

[x] Ten roots per plot were sampled from border rows at R6, gently washed, and rated for root rot severity on a scale of 0–100% on August 9.

[w] Yields were adjusted to 13% moisture after harvesting on October 11 and October 18.

[v] All data were analyzed in SAS 9.4 (SAS Institute, Cary, NC). A generalized linear mixed model analysis of variance was performed using PROC GLIMMIX. Values are least squares means, and values with different letters are significantly different based on least squares means test (α=0.05).

EVALUATION OF SEED TREATMENTS IN SOYBEAN IN CENTRAL INDIANA, 2021 (SOY21-20.ACRE)

S. Shim and D. E. P. Telenko, Department of Botany and Plant Pathology, Purdue University West Lafayette, IN 47907-2054

SOYBEAN (*GLYCINE MAX* AG36XF1)

A trial was established at the Purdue Agronomy Center for Research and Education (ACRE) in Tippecanoe County, Indiana. The experiment was a randomized complete block design with four replications. Plots were 10 feet wide and 30 feet long and consisted of four rows, and the two center rows were used for evaluation. The previous crop was corn. Standard practices for soybean production in Indiana were followed. Soybean cultivar AG36XF1 was planted in 30-inch row spacing at a rate of 8 seeds/foot on May 15. Seed treatments were applied by cooperator. Stand counts were assessed on June 10 and June 22 at V1-V2 and V4 growth stages, respectively. Green stem was visually rated on a scale of 0–100% on October 18. The two center rows of each plot were harvested on October 18, and yields were adjusted to 13% moisture. All data were analyzed in SAS 9.4 (SAS Institute, Cary, NC). A generalized linear mixed model analysis of variance was performed using PROC GLIMMIX. Values are least squares means, and values with different letters are significantly different based on least squares means test (α=0.05).

In 2021, weather conditions were not favorable for soybean disease. There was no significant effect of treatment on stand count, percent of green stem, harvest moisture, test weight, and yield of soybean (Table 14).

TABLE 14. *Effect of Treatment on Stand, Stem Diseases, and Yield of Soybean*

TREATMENT AND RATE/ACRE[z]	STAND COUNT #/A[y] JUN 10	STAND COUNT #/A[y] JUN 22	GREEN % STEM[x] OCT 18	HARVEST MOISTURE %	TEST WEIGHT LB/BU	YIELD[w] BU/ACRE
V-10503 FS 3.78 fl oz/cwt (Zeltera Suite S)	120,879	122,839	33.8	15.0	54.0	74.0
Cruizer Maxx Vibrance 3.22 fl oz/cwt	125,017	122,186	30.0	15.2	54.4	72.7
V-10503 FS 3.78 fl oz/cwt + AVEO EX 0.2 fl oz/cwt	117,176	125,017	47.5	14.6	54.8	64.4
Acceleron 0.8 fl oz/cwt	123,493	120,443	27.5	14.9	54.7	74.8
P-value[v]	0.2528	0.3877	0.4943	0.9075	0.4872	0.3299

[z] Seed treatments were applied by cooperator.

[y] Stand counts were assessed on June 10 and June 22 at V1-V2 and V4 growth stages, respectively.

[x] Percent of green stem was visually rated on a scale of 0–100% on October 18.

[w] Yields were adjusted to 13% moisture after harvesting on October 18.

[v] All data were analyzed in SAS 9.4 (SAS Institute, Cary, NC). A generalized linear mixed model analysis of variance was performed using PROC GLIMMIX. Values are least squares means, and values with different letters are significantly different based on least squares means test (α=0.05).

EVALUATION OF FUNGICIDE PRODUCTS FOR FOLIAR DISEASES IN SOYBEAN IN CENTRAL INDIANA, 2021 (SOY21-20.ACRE)

S. Shim and D. E. P. Telenko, Department of Botany and Plant Pathology, Purdue University West Lafayette, IN 47907-2054

SOYBEAN (*GLYCINE MAX* P35T15E)

Frogeye leaf spot, *Cercospora sojina*
Septoria brown spot, *Septoria glycines*

A trial was established at the Purdue Agronomy Center for Research and Education (ACRE) in Tippecanoe County, Indiana. The experiment was a randomized complete block design with four replications. Plots were 10 feet wide and 30 feet long and consisted of four rows, and the two center rows were used for evaluation. The previous crop was sunflower. Standard practices for soybean production in Indiana were followed. Soybean cultivar P35T15E was planted in 30-inch row spacing at a rate of 140,000 seeds/acre on May 22. Fungicide applications were applied at beginning pod (R3) growth stage on July 27 and were applied at 15 gal/acre at 40 psi using a CO_2 backpack sprayer equipped with a 10-foot boom, fitted with six TJ-VS 8002 nozzles spaced 20 inches apart. Disease ratings were assessed on September 3 at full seed (R6) growth stage. Frogeye leaf spot (FLS) and Septoria brown spot (SBS) were rated for disease severity by visually assessing the percentage of symptomatic leaf area in the upper and lower canopies, respectively. The two center rows of each plot were harvested on October 10, and yields were adjusted to 13% moisture. All data were analyzed in SAS 9.4 (SAS Institute, Cary, NC). A generalized linear mixed model analysis of variance was performed using PROC GLIMMIX. Values are least squares means, and values with different letters are significantly different based on least squares means test (α=0.05).

In 2021, weather conditions were not favorable for soybean disease. FLS and SBS were present in the trial but only remained at low levels. All fungicide treatments reduced FLS and SBS severity over the nontreated control on September 8 (Table 15). There was no significant effect of treatment on percent of green stem, harvest moisture, test weight, and yield of soybean.

TABLE 15. *Effect of Treatment on Foliar Diseases, Stem Diseases, and Yield of Soybean*

TREATMENT AND RATE/ACRE[z]	FLS[y] %	SBS[y] %	GREEN STEM[x] %	HARVEST MOISTURE %	TEST WEIGHT LB/BU	YIELD[w] BU/ACRE
Nontreated control	0.08 a	2.0 a	2.5	14.9	55.2	48.6
Miravis Neo 2.4 SE 13.7 fl oz	0.00 b	0.1 b	15.0	15.1	55.5	55.4
Miravis Top 1.67 SC 13.7 fl oz	0.00 b	0.2 b	20.0	15.4	55.5	49.2
Miravis Neo 2.4 SE 13.7 fl oz + Endigo ZC 4.0 fl oz	0.03 b	0.2 b	25.0	15.3	53.8	56.1
Miravis Top 1.67 SC 13.7 fl oz + Endigo ZC 4.0 fl oz	0.00 b	0.1 b	23.8	15.2	54.3	55.9
Delaro Complete 458 SC 8.0 fl oz	0.00 b	0.3 b	51.3	14.9	55.3	51.7
Priaxor Xemium SC 4.0 fl oz	0.03 b	0.2 b	2.5	15.6	55.4	49.5
Revytek 3.33 LC 8.0 fl oz	0.00 b	0.3 b	22.5	15.0	55.1	56.3
Trivapro 2.21 SE 13.7 fl oz	0.03 b	0.4 b	7.5	15.4	53.8	54.1
P-value[v]	*0.0424*	*0.0001*	*0.0643*	*0.8395*	*0.2828*	*0.2377*

[z] Fungicides were applied on July 27 at beginning pod (R3) growth stage and contained a nonionic surfactant (Preference) at a rate of 0.25% v/v.

[y] Foliar disease incidence was rated on a scale of 0–100% of plants within a plot with disease symptoms on September 3. FLS = frogeye leaf spot, SBS = Septoria brown spot.

[x] Percent of green stem was visually rated on a scale of 0–100% on October 10.

[y] Yields were adjusted to 13% moisture after harvesting on October 10.

[v] All data were analyzed in SAS 9.4 (SAS Institute, Cary, NC). A generalized linear mixed model analysis of variance was performed using PROC GLIMMIX. Values are least squares means, and values with different letters are significantly different based on least squares means test (α=0.05).

XYWAY EFFICACY FOR FOLIAR DISEASES IN SOYBEAN IN CENTRAL INDIANA, 2021 (SOY21-23.ACRE)

S. Shim and D. E. P. Telenko, Department of Botany and Plant Pathology, Purdue University West Lafayette, IN 47907-2054

SOYBEAN (*GLYCINE MAX* P35T15E)

Frogeye leaf spot, *Cercospora sojina*
Septoria brown spot, *Septoria glycines*

A trial was established at the Purdue Agronomy Center for Research and Education (ACRE) in Tippecanoe County, Indiana. The experiment was a randomized complete block design with four replications. Plots were 10 feet wide and 30 feet long and consisted of four rows, and the two center rows were used for evaluation. The previous crop was corn. Standard practices for soybean production in Indiana were followed. Soybean cultivar P35T15E was planted in 30-inch row spacing at a rate of 8 seeds/foot on May 15. Xyway applications were applied in 2x2 application at 10 gal/acre at planting on May 15, and foliar applications were made at V5 and beginning pod (R3) growth stages on July 9 and July 25, respectively. Fungicides were applied at 15 gal/acre at 40 psi using a CO_2 backpack sprayer equipped with a 10-foot boom, fitted with six TJ-VS 8002 nozzles spaced 20 inches apart. Disease ratings were assessed on August 30 at full seed (R6) growth stage. Frogeye leaf spot (FLS) and Septoria brown spot (SBS) were rated for disease severity by visually assessing the percentage of symptomatic leaf area in the upper and lower canopies, respectively. The two center rows of each plot were harvested on October 11, and yields were adjusted to 13% moisture. All data were analyzed in SAS 9.4 (SAS Institute, Cary, NC). A generalized linear mixed model analysis of variance was performed using PROC GLIMMIX. Values are least squares means, and values with different letters are significantly different based on least squares means test ($\alpha=0.05$).

In 2021, weather conditions were not favorable for soybean disease. FLS and SBS were present in the trial but only remained at low levels. All treatments reduced FLS and SBS over the nontreated control on August 30 (Table 16). There was no significant effect of treatment on percent of green stem, harvest moisture, test weight, and yield of soybean.

TABLE 16. *Effect of Treatment on Foliar and Stem Diseases and Yield of Soybean*

TREATMENT, RATE/ACRE, AND TIMING[z]	FLS[y] %	SBS[y] %	GREEN STEM[x] %	HARVEST MOISTURE %	TEST WEIGHT LB/BU	YIELD[w] BU/ACRE
Nontreated control	0.4 a	0.4 a	5.0	15.3	54.5	65.0
Topguard EQ 4.29 5.0 fl oz at R3	0.1 b	0.1 b	3.8	15.3	54.7	63.2
Lucento 4.17 SC 5.0 fl oz at R3	0.0 b	0.0 b	2.8	15.3	55.3	56.2
Miravis Top 1.67 SC 13.7 fl oz at R3	0.0 b	0.1 b	5.0	15.5	55.0	58.2
Revytek 3.33 LC 8.0 fl oz at R3	0.0 b	0.1 b	15.0	15.6	54.8	56.6
Delaro Complete 458 SC 8.0 fl oz at R3	0.1 b	0.1 b	8.8	15.7	54.9	56.8
Xyway LFR 15.2 fl oz 2x2 application	0.1 b	0.1 b	0.0	15.3	55.3	58.8
Topguard EQ 4.29 7.0 fl oz at V5 fb Lucento 4.17 SC 5.0 fl oz at R3	0.0 b	0.1 b	6.3	15.5	55.4	56.5
P-value[v]	0.0005	0.0002	0.0736	0.6366	0.4636	0.4387

[z] Xyway 2x2 application were made at plating on May 15 and fungicide applications on July 9 and July 25 at V5 and R3 growth stages, respectively. All foliar treatments at R3 contained a nonionic surfactant (Preference) at a rate of 0.25% v/v. fb = followed by.

[y] Foliar disease incidence was rated on a scale of 0–100% of plants within a plot with disease symptoms on August 30. FLS = frog-eye leaf spot, SBS = Septoria brown spot.

[x] Percent of green stem was visually rated on a scale of 0–100% on October 11.

[w] Yields were adjusted to 13% moisture after harvesting on October 11.

[v] All data were analyzed in SAS 9.4 (SAS Institute, Cary, NC). A generalized linear mixed model analysis of variance was performed using PROC GLIMMIX. Values are least squares means, and values with different letters are significantly different based on least squares means test (α=0.05).

XYWAY EFFICACY FOR FOLIAR DISEASES IN SOYBEAN IN CENTRAL INDIANA, 2021 (SOY21-24.ACRE)

S. Shim and D. E. P. Telenko, Department of Botany and Plant Pathology, Purdue University West Lafayette, IN 47907-2054

SOYBEAN (*GLYCINE MAX* P35T15E)

Frogeye leaf spot, *Cercospora sojina*
Septoria brown spot, *Septoria glycines*

A trial was established at the Purdue Agronomy Center for Research and Education (ACRE) in Tippecanoe County, Indiana. The experiment was a randomized complete block design with four replications. Plots were 10 feet wide and 30 feet long and consisted of four rows, and the two center rows were used for evaluation. The previous crop was corn. Standard practices for soybean production in Indiana were followed. Soybean cultivar P35T15E was planted in 30-inch row spacing at a rate of 8 seeds/foot on May 15. Xyway applications were applied in-furrow and 2x2 at 10 gal/acre and dribbled by hand at 5 gal/acre on May 15, and foliar applications were made at beginning pod (R3) growth stage on July 27. Foliar fungicides were applied at 15 gal/acre at 40 psi using a CO_2 backpack sprayer equipped with a 10-foot boom, fitted with six TJ-VS 8002 nozzles spaced 20 inches apart. Disease ratings were assessed on August 30 at full seed (R6) growth stage. Frogeye leaf spot (FLS) and Septoria brown spot (SBS) were rated for disease severity by visually assessing the percentage of symptomatic leaf area in the upper and lower canopies, respectively. The two center rows of each plot were harvested on October 11, and yields were adjusted to 13% moisture. All data were analyzed in SAS 9.4 (SAS Institute, Cary, NC). A generalized linear mixed model analysis of variance was performed using PROC GLIMMIX. Values are least squares means, and values with different letters are significantly different based on least squares means test (α=0.05).

In 2021, weather conditions were not favorable for soybean disease. FLS and SBS were present in the trial but only remained at low levels. All treatments reduced FLS and SBS over the nontreated control on August 30 except Xyway at 15.2 fl oz dribbled on FLS and SBS and Xyway 7.6 fl oz 2x2 application on SBS (Table 17). There was no significant effect of treatment on percent of green stem, harvest moisture, test weight, and yield of soybean.

TABLE 17. *Effect of Treatment on Foliar and Stem Diseases and Yield of Soybean*

TREATMENT, RATE/ACRE, AND TIMING[z]	FLS[y] %	SBS[y] %	GREEN STEM[x] %	HARVEST MOISTURE %	TEST WEIGHT LB/BU	YIELD[w] BU/ACRE
Nontreated control	0.5 a	0.9 a	0.0	15.1	55.7	56.4
Xyway LFR 15.2 fl oz in-furrow	0.1 b	0.4 b	0.0	15.4	55.5	51.4
Xyway LFR 7.6 fl oz 2x2 application	0.1 b	0.8 a	0.0	15.0	55.2	52.0
Xyway LFR 15.2 fl oz 2x2 application	0.1 b	0.3 b	0.0	15.2	55.1	56.7
Xyway LFR 15.2 fl oz dribble by hand 10 gal/acre	0.5 a	0.8 a	0.0	15.1	55.1	55.0
Lucento 4.17 SC 5.0 fl oz at R3	0.0 b	0.1 b	0.0	15.5	55.2	55.2
Delaro Complete 458 SC 8.0 fl oz at R3	0.1 b	0.1 b	6.3	15.1	55.1	56.9
P-value[v]	*0.0053*	*0.0001*	*0.4531*	*0.1391*	*0.8609*	*0.5708*

[z] Xyway applications were applied in-furrow, 2x2, and dribbled by hand at 10 gal/acre at planting on May 15, and foliar applications were made at beginning pod (R3) growth stage on July 27.

[y] Foliar disease incidence was rated on a scale of 0–100% of plants within a plot with disease symptoms on August 3 and October 11. FLS = frogeye leaf spot, SBS = Septoria brown spot.

[x] Percent of green stem was visually rated on a scale of 0–100% on October 11.

[w] Yields were adjusted to 13% moisture after harvesting on October 11.

[v] All data were analyzed in SAS 9.4 (SAS Institute, Cary, NC). A generalized linear mixed model analysis of variance was performed using PROC GLIMMIX. Values are least squares means, and values with different letters are significantly different based on least squares means test (α=0.05).

COMPARE THE EFFICACY OF NANO TECHNOLOGY FOR FOLIAR DISEASES IN SOYBEAN IN CENTRAL INDIANA, 2021 (SOY21-26.ACRE)

S. Shim and D. E. P. Telenko, Department of Botany and Plant Pathology, Purdue University West Lafayette, IN 47907-2054

SOYBEAN (*GLYCINE MAX* P35T15E)

Frogeye leaf spot, *Cercospora sojina*
Septoria brown spot, *Septoria glycines*

A trial was established at the Purdue Agronomy Center for Research and Education (ACRE) in Tippecanoe County, Indiana. The experiment was a randomized complete block design with four replications. Plots were 10 feet wide and 30 feet long and consisted of four rows, and the two center rows were used for evaluation. The previous crop was corn. Standard practices for soybean production in Indiana were followed. Soybean cultivar P35T15E was planted in 30-inch row spacing at a rate of 140,000 seeds/acre on May 22. Fungicide applications were applied on July 27 at beginning pod (R3) growth stage and were applied at 15 gal/acre at 40 psi using a CO_2 backpack sprayer equipped with a 10-foot boom, fitted with six TJ-VS 8002 nozzles spaced 20 inches apart. Disease ratings were assessed on September 3 at full seed (R6) growth stage. Frogeye leaf spot (FLS) and Septoria brown spot (SBS) were rated for disease severity by visually assessing the percentage of symptomatic leaf area in the upper and lower canopies, respectively. The two center rows of each plot were harvested on October 10, and yields were adjusted to 13% moisture. All data were analyzed in SAS 9.4 (SAS Institute, Cary, NC). A generalized linear mixed model analysis of variance was performed using PROC GLIMMIX. Values are least squares means, and values with different letters are significantly different based on least squares means test (α=0.05).

In 2021, weather conditions were not favorable for soybean disease. FLS and SBS were present in the trial but only remained at low levels. All treatments that included Miravis Neo reduced SBS over the nontreated control on September 3 (Table 18). There was no significant effect of treatments on FLS severity, harvest moisture, test weight, and yield of soybean.

TABLE 18. *Effect of Treatment on Foliar Disease Severity and Yield of Soybean*

TREATMENT AND RATE/ACRE[z]	FLS[y] %	SBS[y] %	HARVEST MOISTURE %	TEST WEIGHT LB/BU	YIELD[x] BU/ACRE
Nontreated control	0.0	2.4 b	3.8	14.3	55.6
NanoStress 4.0 fl oz	0.0	1.3 bc	0.0	14.2	55.3
NanoPack 4.0 fl oz	0.1	4.0 a	0.0	14.6	55.5
NanoN 4.0 fl oz	0.2	1.1 bc	0.0	14.4	55.2
Miravis Neo 2.5 SE 13.7 fl oz	0.0	0.3 c	2.5	14.5	55.2
Miravis Neo 2.5 SE 13.7 fl oz + NanoStress 4.0 fl oz	0.0	0.3 c	6.3	14.3	54.9
Miravis Neo 2.5 SE 13.7 fl oz + NanoPack 4.0 fl oz	0.0	0.1 c	7.5	14.5	55.8
Miravis Neo 2.5 SE 13.7 fl oz + NanoN 4.0 fl oz	0.0	0.3 c	1.3	14.4	54.8
Miravis Neo 2.5 SE 13.7 fl oz + NanoPro 4.0 fl oz	0.0	0.1 c	6.3	14.2	54.9
P-value[w]	0.3494	0.0002	0.7685	0.6656	0.4926

[z] Fungicide applications were made on July 27 at beginning pod (R3) growth stage.

[y] Foliar disease incidence was rated on a scale of 0–100% of plants within a plot with disease symptoms on September 3. FLS = frogeye leaf spot, SBS = Septoria brown spot.

[x] Yields were adjusted to 13% moisture after harvesting on October 10.

[w] All data were analyzed in SAS 9.4 (SAS Institute, Cary, NC). A generalized linear mixed model analysis of variance was performed using PROC GLIMMIX. Values are least squares means, and values with different letters are significantly different based on least squares means test (α=0.05).

TEMPERA EFFICACY FOR DISEASE IN SOYBEAN IN CENTRAL INDIANA, 2021 (SOY21-28.ACRE)

S. Shim and D. E. P. Telenko, Department of Botany and Plant Pathology, Purdue University, West Lafayette, IN 47907-2054

SOYBEAN (*GLYCINE MAX* P35T15E)

Frogeye leaf spot, *Cercospora sojina*
Septoria brown spot, *Septoria glycines*

A trial was established at the Purdue Agronomy Center for Research and Education (ACRE) in Tippecanoe County, Indiana. The experiment was a randomized complete block design with four replications. Plots were 10 feet wide and 30 feet long and consisted of four rows, and the two center rows were used for evaluation. The previous crop was corn. Standard practices for soybean production in Indiana were followed. Soybean cultivar P35T15E was planted in 30-inch row spacing at a rate of 8 seeds/foot on May 15. Tempera applications were applied in-furrow at 10 gal/acre at planting on May 15. Disease ratings were assessed on September 3 at full seed (R6) growth stage. Frogeye leaf spot (FLS) and Septoria brown spot (SBS) were rated for disease severity by visually assessing the percentage of symptomatic leaf area in the upper and lower canopies, respectively. The two center rows of each plot were harvested on October 11, and yields were adjusted to 13% moisture. All data were analyzed in SAS 9.4 (SAS Institute, Cary, NC). A generalized linear mixed model analysis of variance was performed using PROC GLIMMIX. Values are least squares means, and values with different letters are significantly different based on least squares means test (α=0.05).

In 2021, weather conditions were not favorable for soybean disease. FLS and SBS were present in the trial but only remained at low levels. There was no significant effect of treatment on FLS and SBS severity, harvest moisture, test weight, and yield of soybean (Table 19).

TABLE 19. *Effect of Treatment on Foliar Disease Severity and Yield of Soybean*

TREATMENT AND RATE/ACRE[z]	FLS[y] %	SBS[y] %	HARVEST MOISTURE %	TEST WEIGHT LB/BU	YIELD[x] BU/ACRE
Nontreated control	0.5	0.7	14.9	54.2	72.6
Tempera Plus 5.4 fl oz in-furrow	0.5	0.6	14.9	54.8	73.2
P-value[w]	—	0.9037	0.8425	0.1216	0.4292

[z] Tempera in-furrow application was made at plating on May 15.

[y] Foliar disease incidence was rated on a scale of 0–100% of plants within a plot with disease symptoms on September 3. FLS = frogeye leaf spot, SBS = Septoria brown spot.

[x] Yields were adjusted to 13% moisture after harvesting on October 11.

[w] All data were analyzed in SAS 9.4 (SAS Institute, Cary, NC). A generalized linear mixed model analysis of variance was performed using PROC GLIMMIX. Values are least squares means, and values with different letters are significantly different based on least squares means test (α=0.05).

EVALUATION OF OROAGRI PRODUCTS FOR WHITE MOLD IN SOYBEAN IN CENTRAL INDIANA, 2021 (SOY21-29.ACRE)

S. Shim and D. E. P. Telenko, Department of Botany and Plant Pathology, Purdue University, West Lafayette, IN 47907-2054

SOYBEAN (*GLYCINE MAX* P35T15E)

White mold, *Sclerotinia sclerotiorum*
Frogeye leaf spot, *Cercospora sojina*
Septoria brown spot, *Septoria glycines*

A trial was established at the Purdue Agronomy Center for Research and Education (ACRE) in Tippecanoe County, Indiana. The experiment was a randomized complete block design with four replications. Plots were 6.7 feet wide and 30 feet long and consisted of four rows, and the two center rows were used for evaluation. The previous crop was sunflower. Standard practices for soybean production in Indiana were followed. Soybean cultivar P35T15E was planted in 20-inch row spacing at a rate of 8 seeds/foot on May 15. Inoculum of *S. sclerotiorum* was applied on the seedbed at 1.25 g/foot at planting. Preemergence (PRE-E) pesticide applications were applied at 20 gal/acre and at beginning bloom (R1) growth stage were applied at 15 gal/acre at 40 psi using a CO_2 backpack sprayer equipped with a 10-foot boom, fitted with six TJ-VS 8002 nozzles spaced 20 inches apart. (PRE-E) application were applied on June 15 and fungicide applications on July 13 at beginning bloom (R1) growth stage. Disease ratings were assessed on August 30 at full seed (R6) growth stage. Frogeye leaf spot (FLS) and Septoria brown spot (SBS) were rated for disease severity by visually assessing the percentage of symptomatic leaf area in the upper and lower canopies, respectively. White mold disease was assessed by counting the number of plants in each plot with symptoms. The two center rows of each plot were harvested on October 10, and yields were adjusted to 13% moisture. All data were analyzed in SAS 9.4 (SAS Institute, Cary, NC). A generalized linear mixed model analysis of variance was performed using PROC GLIMMIX. Values are least squares means, and values with different letters are significantly different based on least squares means test (α=0.05).

In 2021, weather conditions were not favorable for soybean disease. White mold, FLS, and SBS were present in the trial but only remained at low levels. SBS was reduced mostly by Contans plus Valor XLT on August 30 over the nontreated control (Table 20). There were no significant differences between treatments and the nontreated control for FLS and white mold on August 30. There was no significant effect of treatment on harvest moisture, test weight, and yield of soybean.

TABLE 20. *Effect of Fungicide on Foliar Diseases Severity and Yield of Soybean*

TREATMENT, RATE/ACRE, AND TIMING[z]	FLS[y] %	SBS[y] %	WHITE MOLD[x] #/PLOT	HARVEST MOISTURE %	TEST WEIGHT LB/BU	YIELD[w] BU/ACRE
Nontreated control—Valor XLT 4.0 oz PRE-E	0.5	0.5 a	0.0	13.6	55.4	67.4
OR-079-B 2.0 pts + Valor XLT 4.0 oz PRE-E	0.5	0.5 a	0.3	13.7	55.6	67.4
OR-369-A 2.0 pts + Valor XLT 4.0 oz PRE-E	0.4	0.5 a	0.3	13.5	55.3	67.1
OR 009-A 2.0 pts + Valor XLT 4.0 oz PRE-E	0.4	0.5 a	0.0	13.6	55.5	68.3
Contans WG 2.0 lbs + Valor XLT 4.0 oz PRE-E	0.4	0.3 b	0.5	13.7	55.1	70.4
Valor XLT 4.0 oz PRE-E fb OR 009-A 1.0 pt at R1	0.5	0.5 a	0.0	13.6	55.2	69.7
OR-079-B 2.0 pts + Valor XLT 4.0 oz PRE-E fb OR 009-A 1.0 pt at R1	0.4	0.5 a	0.0	13.6	55.5	67.4
Endura 70 WDG 8.0 oz at R1	0.2	0.5 a	0.0	13.5	55.2	67.0
P-value[v]	*0.5521*	*0.0239*	*0.5962*	*0.9474*	*0.3301*	*0.6279*

[z] Preemergence (PRE-E) application were applied on June 15, and fungicide applications were applied on July 13 at R1 growth stage. fb = followed by. All plots inoculated with *S. sclerotiorum*.

[y] Foliar disease incidence was rated on a scale of 0–100% of plants within a plot with disease symptoms on August 30. FLS = frogeye leaf spot, SBS = Septoria brown spot.

[x] White mold disease was assessed by counting the number of plants in plots with symptoms on August 30.

[w] Yields were adjusted to 13% moisture after harvesting on October 10.

[v] All data were analyzed in SAS 9.4 (SAS Institute, Cary, NC). A generalized linear mixed model analysis of variance was performed using PROC GLIMMIX. Values are least squares means, and values with different letters are significantly different based on least squares means test (α=0.05).

FUNGICIDE COMPARISON OF OROAGRI PRODUCTIONS IN SOYBEAN IN CENTRAL INDIANA, 2021 (SOY21-30.ACRE)

S. Shim and D. E. P. Telenko, Department of Botany and Plant Pathology, Purdue University West Lafayette, IN 47907-2054

SOYBEAN (*GLYCINE MAX* P35T15E)

Frogeye leaf spot, *Cercospora sojina*
Septoria brown spot, *Septoria glycines*

A trial was established at the Purdue Agronomy Center for Research and Education (ACRE) in Tippecanoe County, Indiana. The experiment was a randomized complete block design with four replications. Plots were 10 feet wide and 30 feet long and consisted of four rows, and the two center rows were used for evaluation. The previous crop was corn. Standard practices for soybean production in Indiana were followed. Soybean cultivar P35T15E was planted in 30-inch row spacing at a rate of 140,000 seeds/acre on May 22. Fungicide applications were applied on July 31 at beginning pod (R3) growth stage and were applied at 15 gal/acre at 40 psi using a CO_2 backpack sprayer equipped with a 10-foot boom, fitted with six TJ-VS 8002 nozzles spaced 20 inches apart. Disease ratings were assessed on September 3 at full seed (R6) growth stage. Frogeye leaf spot (FLS) and Septoria brown spot (SBS) were rated for disease severity by visually assessing the percentage of symptomatic leaf area in the upper and lower canopies, respectively. The two center rows of each plot were harvested on October 10, and yields were adjusted to 13% moisture. All data were analyzed in SAS 9.4 (SAS Institute, Cary, NC). A generalized linear mixed model analysis of variance was performed using PROC GLIMMIX. Values are least squares means, and values with different letters are significantly different based on least squares means test (α=0.05).

In 2021, weather conditions were not favorable for soybean disease. FLS and SBS were present in the trial but only remained at low levels. All fungicides reduced SBS over the nontreated control on September 3 (Table 21). There was no significant effect of treatment on FLS severity, harvest moisture, test weight, and yield of soybean.

TABLE 21. *Effect of Treatment on Foliar Diseases and Yield of Soybean*

TREATMENT AND RATE/ACRE[z]	FLS[y] %	SBS[y] %	HARVEST MOISTURE %	TEST WEIGHT LB/BU	YIELD[x] BU/ACRE
Nontreated control	0.1	3.0 a	14.8	54.8	47.1
Topguard EQ 4.29 5.0 fl oz	0.0	0.5 b	14.8	55.6	49.3
Topguard EQ 4.29 5.0 fl oz + OR-099-E 0.25% v/v	0.0	0.9 b	14.6	55.9	48.2
Topguard EQ 4.29 5.0 fl oz + OR-295-A 0.25% v/v	0.0	0.7 b	15.3	55.4	50.0
Topguard EQ 4.29 5.0 fl oz + OR-025-F 1.0 pt/A	0.0	0.9 b	15.5	55.9	46.7
Topguard EQ 4.29 5.0 fl oz + OR-009-A 0.4 % v/v	0.0	0.6 b	15.4	55.1	44.8
P-value[w]	0.2978	0.0360	0.4136	0.2832	0.7229

[z] Fungicide applications were made on July 31 at beginning pod (R3) growth stage.

[y] Foliar disease incidence was rated on a scale of 0–100% of plants within a plot with disease symptoms on September 3. FLS = frogeye leaf spot, SBS = Septoria brown spot.

[x] Yields were adjusted to 13% moisture after harvesting on October 10.

[w] All data were analyzed in SAS 9.4 (SAS Institute, Cary, NC). A generalized linear mixed model analysis of variance was performed using PROC GLIMMIX. Values are least squares means, and values with different letters are significantly different based on least squares means test (α=0.05).

EVALUATION OF FOLIAR FUNGICIDES AND ORGANIC VARIETIES FOR SCAB MANAGEMENT IN CENTRAL INDIANA, 2021 (WHT21-01.ACRE)

C. R. Da Silva, S. Shim, and D. E. P. Telenko, Department of Botany and Plant Pathology, Purdue University West Lafayette, IN 47907-2054

WHEAT (*TRITICUM AESTIVUM* KASKASKIA AND HARPOON)

Fusarium head blight, *Fusarium graminearum*

A trial was established at the Purdue Agronomy Center for Research and Education (ACRE) in Tippecanoe County, Indiana. The experiment was a randomized complete block design with four replications. Plots were 7.5 feet wide and 20 feet long and consisted of 12 rows spaced 7.5 inches apart, and the center of each plot was used for evaluation. The previous crop was corn. Organic wheat varieties Kaskaskia and Harpoon were planted in 7.5-inch row spacing using a drill on October 14, 2020. All fungicide applications were applied at 15 gal/acre and 40 psi using a CO_2 backpack sprayer equipped with a 10-foot boom, fitted with six TJ-VS 8002 nozzles spaced 20 inches apart and directed forward and backward at a 45-degree angle. Fungicides were applied on May 22 and May 23, 2021, at Feekes growth stage 10.5.1. All plots were inoculated with a mixture of isolates of *Fusarium graminearum* endemic to Indiana on May 23 and May 24, 2021, with a spore suspension (50,000 spores/ml) applied at 300 ml/plot. Disease ratings were assessed on June 11. Fusarium head blight (FHB) incidence was measured as the number of infected heads out of 60 plants in each plot and calculated as a percentage. FHB severity was rated by visually assessing the percentage of the infected head, and FHB index was calculated as (% FHB incidence multiplied by average FHB severity)/100 per plot. The eight center rows of each plot were harvested with a Kincaid plot combine on July 7, and yields were adjusted to 13.5% moisture. All data were analyzed in SAS 9.4 (SAS Institute, Cary, NC). A generalized linear mixed model analysis of variance was performed using PROC GLIMMIX. Values are least squares means, and values with different letters are significantly different based on least squares means test (α=0.05).

In 2021, weather conditions were not favorable for FHB. No differences were detected between treatments for FHB incidence, severity, and index and the nontreated control on June 11 (Table 22). The percent of Fusarium damaged kernels (FDK) was lowest in the Kaskaskia cultivar and when treated with Prosaro and Actinovate. The concentration of deoxynivalenol (DON) was lowest in the cultivar Kaskaskia. An application of Pacesetter increased DON over the nontreated control. There was no difference in treatment for yield of wheat.

TABLE 22. *Effect of Cultivar and Fungicide on Fusarium Head Blight (FBH) and Foliar Diseases in Organic Wheat*

CULTIVAR, TREATMENT, AND RATE/ACRE[z]	FHB % DI[y]	FHB % DS[x]	FHB INDEX[w]	FDK[v] %	DON[u] PPM	YIELD[t] BU/ACRE
Cultivar						
Kaskaskia	17.6[s]	3.8	0.6	14.5 b	0.067 b[v]	41.4
Harpoon	20.3	4.5	0.9	20.8 a	0.341 a	46.4
Fungicide programs						
Nontreated control	21.5	6.3	1.4	20.7 a	0.150 b	41.7
Prosaro 421 SC 8.2 fl oz	23.3	2.5	0.6	14.7 c	0.243 ab	45.2
ChampION 50 WP 1.5 lb	17.1	1.9	0.4	18.6 ab	0.200 ab	46.6
Pacesetter WS 13.0 fl oz	17.7	3.3	0.6	18.2 ab	0.367 a	43.2
Sonata 1.0 qt	14.9	7.3	0.8	18.3 ab	0.120 b	44.0
Actinovate AG 12.0 fl oz	20.2	3.5	0.8	16.1 bc	0.150 b	42.9
P-value cultivar[u]	*0.2606*	*0.6373*	*0.1798*	*0.0001*	*0.0001*	*0.1880*
P-value fungicide	*0.2389*	*0.3333*	*0.1916*	*0.0223*	*0.0957*	*0.9796*
*P-value cultivar*fungicide*	*0.1083*	*0.8776*	*0.4629*	*0.7277*	*0.1323*	*0.8851*

[z] Fungicides were applied on May 22 and May 23, 2021, at Feekes growth stage 10.5.1. All plots were inoculated with a mixture of isolates of *Fusarium graminearum* endemic to Indiana, with a spore suspension (50,000 spores/ml) applied at 300 ml/plot on May 23 and May 24.

[y] FHB disease incidence (DI) was measured as the number of infected heads out of 60 plants in each plot and calculated as a percentage on June 11.

[x] FHB disease severity (DS) was visually assessed as a percentage of the infected head. FHB = Fusarium head blight on June 11.

[w] FHB index was calculated as (% FHB incidence multiplied by average FHB severity)/100 per plot on June 11.

[v] Fusarium damaged kernels (FDK) were visually assessed as a percentage of Fusarium damaged heads on September 21.

[u] Analysis of the mycotoxin deoxynivalenol (DON) was completed by the University of Minnesota DON Testing Lab on October 21.

[t] Yields were adjusted to 13.5% moisture after harvesting on July 7.

[s] All data were analyzed in SAS 9.4 (SAS Institute, Cary, NC). A generalized linear mixed model analysis of variance was performed using PROC GLIMMIX. Values are least squares means, and values with different letters are significantly different based on least squares means test (α=0.05).

EVALUATION OF FOLIAR FUNGICIDES FOR SCAB MANAGEMENT IN CENTRAL INDIANA, 2021 (WHT21-02.ACRE)

S. Shim and D. E. P. Telenko, Department of Botany and Plant Pathology, Purdue University, West Lafayette, IN 47907-2054

WHEAT (*TRITICUM AESTIVUM* P25R40)

Fusarium head blight, *Fusarium graminearum*

A trial was established at the Purdue Agronomy Center for Research and Education (ACRE) in Tippecanoe County, Indiana. The experiment was a randomized complete block design with four replications. Plots were 7.5 feet wide and 20 feet long and consisted of 12 rows spaced 7.5 inches apart, and the center of each plot was used for evaluation. The previous crop was corn. Prior to planting, the field was disked and chisel-plowed on October 10, 2020. Nitrogen (28%) at 30 gal/acre was applied on March 10, 2020. On October 16, 2020, wheat cultivar P25R40 was drilled at 7.5-inch spacing. All fungicide applications were applied at 15 gal/acre and 40 psi using a CO2 backpack sprayer equipped with a 10-foot boom, fitted with six TJ-VS 8002 nozzles spaced 20 inches apart and directed forward and backward at a 45-degree angle. Fungicides were applied on May 20, May 22, and May 29, 2021, at Feekes growth stages 10.3, 10.5.1, and 10.5.1 + 6 days, respectively. All plots were inoculated with a mixture of isolates of *Fusarium graminearum* endemic to Indiana on May 22. The spore suspension (50,000 spores/ml) was applied at 300 ml/plot with a CO_2 handheld sprayer. Disease ratings were assessed on June 11, 2021. Fusarium head blight (FHB) incidence was measured as the number of infected heads out of 60 plants in each plot and calculated as a percentage. FHB severity was rated by visually assessing the percentage of the infected head. FHB index was calculated as (% FHB incidence multiplied by average FHB severity)/100 per plot. The eight center rows of each plot were harvested with a Kincaid plot combine on July 7, and yields were adjusted to 13.5% moisture. All data were analyzed in SAS 9.4 (SAS Institute, Cary, NC). A generalized linear mixed model analysis of variance was performed using PROC GLIMMIX. Values are least squares means, and values with different letters are significantly different based on least squares means test (α=0.05).

In 2021, weather conditions were not favorable for FHB. FHB incidence was reduced by all fungicides over the nontreated control on June 11 except for Caramba applied at 10.5.1 (Table 23). No differences were detected for FHB index and severity as compared to the nontreated control. The concentration of deoxynivalenol (DON) was reduced over the nontreated control in all treatments except Prosaro applied at 10.5.1 and Miravis Ace applied at 10.3. Fusarium damaged kernels (FDK) were reduced in all treatments over the nontreated control except for Caramba and Sphaerex applied at 10.5.1. There were no significant differences in yield.

TABLE 23. *Effect of Fungicide on Fusarium Head Blight (FHB) and Foliar Diseases in Wheat*

TREATMENT AND RATE/ACRE[z]	FHB % DI[y]	FHB % DS[x]	FHB INDEX[w]	FDK[v] %	DON[u] PPM	YIELD[t] BU/ACRE
Nontreated control	25.8 a[s]	6.5	2.2	0.925 a	12.0 a	98.2
Prosaro 421 SC 6.5 fl oz at 10.5.1	15.0 bc	2.1	0.4	0.758 ab	8.5 bcd	86.8
Caramba 90 EC 13.5 fl oz at 10.5.1	19.6 ab	1.2	0.2	0.573 bc	11.3 ab	97.2
Sphaerex 7.3 fl oz at 10.5.1	14.2 bc	2.1	0.3	0.410 cd	9.5 abc	90.1
Miravis Ace 5.2 SC 13.7 fl oz at 10.3	9.6 c	5.2	0.5	0.713 ab	7.3 cde	101.8
Miravis Ace 5.2 SC 13.7 fl oz at 10.5.1	10.4 c	2.9	0.4	0.385 ab	6.5 cde	95.3
Miravis Ace 5.2 SC 13.7 fl oz at 10.5.1+ 4 days	14.2 bc	3.3	0.5	0.370 cd	5.8 de	86.3
Miravis Ace 5.2 SC 13.7 fl oz at 10.5.1 fb Prosaro 421 SC 6.5 fl oz at 10.5.1 + 4 days	7.1 c	2.1	0.2	0.213 cd	6.3 de	88.6
Miravis Ace 5.2 SC 13.7 fl oz at 10.5.1 fb Caramba 90 EC 13.5 fl oz 10.5.1 + 4 days	7.5 c	1.5	0.1	0.173 d	5.8 de	89.0
Miravis Ace 5.2 SC 13.7 fl oz at 10.5.1 fb Folicur 3.6 F 4.0 fl oz at 10.51 + 4 days	10.8 bc	3.2	0.3	0.248 d	4.3 e	88.7
P-value[s]	0.0066	0.1838	0.1826	0.0001	0.0001	0.4829

[z] Fungicide treatments were applied on May 20, May 22, and May 29, 2021, at Feekes growth stages 10.3, 10.5.1, and 10.5.1 + 6 days, respectively. All treatments contained a nonionic surfactant (Preference) at a rate of 0.125% v/v. All plots were inoculated with *Fusarium graminearum* spore suspension (50,000 spores/ml) after the treatment at Feekes 10.5.1. Spore suspension was applied at 300 ml/plot with a handheld sprayer on May 23. fb = followed by.

[y] FHB disease incidence (DI) was measured as the number of infected heads out of 60 plants in each plot and calculated as a percentage on June 11.

[x] FHB disease severity (DS) was visually assessed as a percentage of the infected head on June 11. FHB = Fusarium head blight.

[w] FHB index was calculated as (% FHB incidence multiplied by average FHB severity)/100 per plot on June 11.

[v] Fusarium damaged kernels (FDK) were visually assessed as a percentage of Fusarium damaged heads on September 21.

[u] Analysis of the mycotoxin deoxynivalenol (DON) was completed by the University of Minnesota DON Testing Lab on October 21.

[t] Yields were adjusted to 13.5% moisture after harvesting on July 7.

[s] All data were analyzed in SAS 9.4 (SAS Institute, Cary, NC). A generalized linear mixed model analysis of variance was performed using PROC GLIMMIX. Values are least squares means, and values with different letters are significantly different based on least squares means test (α=0.05).

EVALUATION OF FOLIAR FUNGICIDES AND VARIETIES FOR SCAB MANAGEMENT IN CENTRAL INDIANA, 2021 (WHT21-03.ACRE)

S. Shim and D. E. P. Telenko, Department of Botany and Plant Pathology, Purdue University, West Lafayette, IN 47907-2054

WHEAT (*TRITICUM AESTIVUM* P25R40 AND P25R61)

Fusarium head blight, *Fusarium graminearum*

A trial was established at the Purdue Agronomy Center for Research and Education (ACRE) in Tippecanoe County, Indiana. The experiment was a randomized complete block design with four replications. Plots were 7.5 feet wide and 20 feet long and consisted of 12 rows spaced 7.5 inches apart, and the center of each plot was used for evaluation. The previous crop was corn. Prior to planting, the field was disked and chisel-plowed on October 10, 2020. Nitrogen (28%) at 30 gal/acre was applied on March 10, 2020. On October 16, 2020, wheat cultivar P25R40 was drilled at 7.5 inches spacing. All fungicide applications were applied at 15 gal/acre and 40 psi using a CO_2 backpack sprayer equipped with a 10-feet boom, fitted with six TJ-VS 8002 nozzles spaced 20 inches apart and directed forward and backward at a 45-degree angle. Fungicides were applied on May 20, May 22, and May 29, 2021, at Feekes growth stages 10.3, 10.5.1 and 10.5.1 + 6 days, respectively. All plots were inoculated with a mixture of isolates of *Fusarium graminearum* endemic to Indiana on May 22. The spore suspension (50,000 spores/ml) was applied at 300 ml/plot with a CO_2 handheld sprayer. Disease ratings were assessed on June 11, 2021. Fusarium head blight (FHB) incidence was measured as the number of infected heads out of 60 plants in each plot and calculated as a percentage. FHB severity was rated by visually assessing the percentage of the infected head, and FHB index was calculated as (% FHB incidence multiplied by average FHB severity)/100 per plot. The eight center rows of each plot were harvested with a Kincaid plot combine on July 7, and yields were adjusted to 13.5% moisture. All data were analyzed in SAS 9.4 (SAS Institute, Cary, NC). A generalized linear mixed model analysis of variance was performed using PROC GLIMMIX. Values are least squares means, and values with different letters are significantly different based on least squares means test (α=0.05).

In 2021, weather conditions were not favorable for FHB. There was a significant interaction between cultivar and treatment (>0.05); therefore, treatment affect was evaluated across each cultivar (Table 24). In the susceptible cultivar, P25R40, FHB incidence was reduced by Miravis Ace applied at 10.5.1 and 10.3 and at 10.5.1 followed by (fb) Folicur at 10.5.1+6 days. All fungicides reduced FHB severity when compared to nontreated, inoculated control but not the nontreated noninoculated control in P25R40. FHB index was lowest with the Miravis Ace fb Folicur but was not different from the single Miravis Ace applications for P25R40. In P25R40, deoxynivalenol (DON) was reduced by Prosaro and Miravis Ace fb Folicur, while Fusarium damaged kernels (FDK) were lowest with Miravis Ace applied at 10.5.1 and Miravis Ace fb Folicur treatments over the nontreated controls. There was no difference in fungicide treatments for FHB incidence, severity, index, DON, and FDK in the resistant cultivar P25R61. No differences in yield were detected in either cultivar.

TABLE 24. *Effect of Cultivar and Fungicide on Fusarium Head Blight (FHB), Deoxynivalenol (DON), Fusarium Damaged Kernels (FDK), and Yield of Wheat*

TREATMENT AND RATE/ACRE[z]	FHB % DI[y]	FHB % DS[x]	FHB INDEX[w]	DON[v] PPM	FDK[u] %	YIELD[t] BU/ACRE
P25R40						
Nontreated control, inoculated control	38.7 a	5.4 a	2.0 a	0.89 ab	10.5 ab	98.9
Nontreated, noninoculated control	27.1 ab	2.2 bc	0.6 bc	0.99 a	11.8 a	92.9
Prosaro 421 SC 6.5 fl oz at 10.5.1	26.3 ab	2.4 bc	0.7 b	0.44 c	7.0 bc	91.6
Miravis Ace 5.2 SC 13.7 fl oz at 10.5.1	13.8 b	1.8 bc	0.2 bc	0.48 bc	4.4 c	87.2
Miravis Ace 5.2 SC 13.7 fl oz at 10.3	15.4 b	3.0 bc	0.4 bc	0.67 abc	6.8 bc	97.6
Miravis Ace 13.7 fl oz at 10.5.1 fb Folicur 3.6 F 4.0 fl oz at 10.5.1 + 6d	12.1 b	1.3 c	0.2 c	0.34 c	4.4 c	106.8
P-value[s]	*0.0126*	*0.0004*	*0.0001*	*0.0325*	*0.0114*	*0.0652*
P25R61						
Nontreated control, inoculated control	17.1	1.8	0.3	0.05	9.5	92.9
Nontreated, noninoculated control	18.7	1.8	0.4	0.07	9.3	93.0
Prosaro 421 SC 6.5 fl oz at 10.5.1	12.9	1.9	0.3	0.00	7.0	94.7
Miravis Ace 5.2 SC 13.7 fl oz at 10.5.1	9.2	1.8	0.1	0.00	5.8	95.0
Miravis Ace 5.2 SC 13.7 fl oz at 10.3	8.8	1.5	0.1	0.03	5.0	95.7
Miravis Ace 13.7 fl oz at 10.5.1 fb Folicur 3.6 F 4.0 fl oz at 10.5.1 + 6d	8.8	1.7	0.2	0.00	5.5	91.7
P-value[s]	*0.1274*	*0.9874*	*0.438*	*0.0519*	*0.0654*	*0.9019*

[z] Fungicide treatments were applied on May 20, May 22, and May 29, 2021, at Feekes growth stages 10.3, 10.5.1, and 10.5.1 + 6 days, respectively. All treatments contained a nonionic surfactant (Preference) at a rate of 0.125% v/v. All plots inoculated with *Fusarium graminearum* spore suspension (50,000 spores/ml) after the treatment at Feekes 10.5.1. Spore suspension applied at 300 ml/plot with handheld sprayer on May 23. fb = followed by.

[y] FHB incidence was measured as the number of infected heads out of 60 plants in each plot and calculated as a percentage.

[x] FHB severity was visually assessed as a percentage of the infected head. FHB = Fusarium head blight.

[w] FHB index was calculated as (% FHB incidence multiplied by average FHB severity)/100 per plot.

[v] Analysis of the mycotoxin deoxynivalenol (DON) was completed by the University of Minnesota DON Testing Lab.

[u] Fusarium damaged kernels (FDK) were visually assessed as a percentage of Fusarium damaged heads.

[t] Yields were adjusted to 13.5% moisture after harvesting on July 7.

[s] All data were analyzed in SAS 9.4 (SAS Institute, Cary, NC). A generalized linear mixed model analysis of variance was performed using PROC GLIMMIX. Values are least squares means, and values with different letters are significantly different based on least squares means test (α=0.05).

EVALUATION OF FOLIAR FUNGICIDES FOR WHEAT DISEASE MANAGEMENT IN CENTRAL INDIANA, 2021 (WHT21-06.ACRE)

S. Shim and D. E. P. Telenko, Department of Botany and Plant Pathology, Purdue University West Lafayette, IN 47907-2054

WHEAT (*TRITICUM AESTIVUM* P25R40)

Fusarium head blight, *Fusarium graminearum*

A trial was established at the Purdue Agronomy Center for Research and Education (ACRE) in Tippecanoe County, Indiana. The experiment was a randomized complete block design with four replications. Plots were 7.5 feet wide and 20 feet long and consisted of 12 rows spaced 7.5 inches apart, and the center of each plot was used for evaluation. The previous crop was corn. Prior to planting, the field was disked and chisel-plowed on October 10, 2020. Nitrogen (28%) at 30 gal/acre was applied on March 10, 2020. On October 16, 2020, wheat cultivar P25R40 was drilled at 7.5-inch spacing. All fungicide applications were applied at 15 gal/acre and 40 psi using a CO2 backpack sprayer equipped with a 10-foot boom, fitted with six TJ-VS 8002 nozzles spaced 20 inches apart and directed forward and backward at a 45-degree angle. Fungicides were applied on May 22, 2021, at Feekes growth stage 10.5.1. All plots were inoculated with a mixture of isolates of *Fusarium graminearum* endemic to Indiana on May 22. The spore suspension (50,000 spores/ml) was applied at 300 ml/plot with a CO_2 handheld sprayer. Disease ratings were assessed on June 11, 2021. Fusarium head blight (FHB) incidence was measured as the number of infected heads out of 60 plants in each plot and calculated as a percentage. FHB severity was rated by visually assessing the percentage of the infected head, and FHB index was calculated as (% FHB incidence multiplied by average FHB severity)/100 per plot. The eight center rows of each plot were harvested with a Kincaid plot combine on July 7, and yields were adjusted to 13.5% moisture. All data were analyzed in SAS 9.4 (SAS Institute, Cary, NC). A generalized linear mixed model analysis of variance was performed using PROC GLIMMIX. Values are least squares means, and values with different letters are significantly different based on least squares means test ($\alpha=0.05$).

In 2021, weather conditions were not favorable for FHB. FHB incidence and index were reduced by all fungicides over the nontreated control on June 11 (Table 25). FHB severity and the concentration of deoxynivalenol (DON) was not significantly reduced over the nontreated control for all treatments. There was no difference in wheat yield.

TABLE 25. *Effect of Fungicide on Fusarium Head Blight (FHB), Deoxynivalenol (DON), Fusarium Damaged Kernels (FDK), and Yield of Wheat*

TREATMENT AND RATE/ACRE[z]	FHB % DI[y]	FHB % DS[x]	FHB INDEX[w]	DON[v] PPM	FDK[u] %	YIELD[t] BU/ACRE
Nontreated control	37.1 a	7.1	2.6 a	0.86	10.8 a	90.5
Prosaro 421 SC 8.2 fl oz	16.7 b	3.1	0.5 b	0.68	9.0 a	94.6
Prosaro Pro SC 10.3 fl oz	12.5 b	3.1	0.5 b	0.64	8.3 ab	102.5
Miravis Ace 5.2 SC 13.7 fl oz	11.7 b	1.6	0.2 b	0.35	5.5 b	86.2
P-value[s]	0.0060	0.1154	0.0413	0.0571	0.0300	0.2736

[z] Fungicide treatments applied at Feekes 10.5.1 all treatments contained a nonionic surfactant (Preference) at a rate of 0.125% v/v. All plots were inoculated with *Fusarium graminearum* spore suspension (50,000 spores/ml) after the treatment at Feekes 10.5.1. Spore suspension was applied at 300 ml/plot with handheld sprayer on May 23.

[y] FHB incidence was measured as the number of infected heads out of 60 plants in each plot and calculated as a percentage.

[x] FHB severity was visually assessed as a percentage of the infected head. FHB = Fusarium head blight.

[w] FHB index was calculated as (% FHB incidence multiplied by average FHB severity)/100 per plot.

[v] Analysis of the mycotoxin deoxynivalenol (DON) was completed by the University of Minnesota DON Testing Lab.

[u] Fusarium damaged kernels (FDK) were visually assessed as a percentage of Fusarium damaged heads.

[t] Yields were adjusted to 13.5% moisture after harvesting on July 7.

[s] All data were analyzed in SAS 9.4 (SAS Institute, Cary, NC). A generalized linear mixed model analysis of variance was performed using PROC GLIMMIX. Values are least squares means followed by standard errors. Values with different letters are significantly different based on least squares means test (α=0.05).

PINNEY PURDUE AGRICULTURAL CENTER (PPAC)

UNIFORM FUNGICIDE COMPARISON FOR TAR SPOT IN CORN IN NORTHWESTERN INDIANA, 2021 (COR21-02_UFTTAR.PPAC)

T. J. Ross, S. Shim, and D. E. P. Telenko, Department of Botany and Plant Pathology, Purdue University, West Lafayette, IN 47907-2054

CORN (*ZEA MAYS* W2585SSRIB)

Tar spot, *Phyllachora maydis*

A trial was established at the Pinney Purdue Agricultural Center (PPAC) in Porter County, Indiana. The experiment was a randomized complete block design with four replications. Plots were 10 feet wide and 30 feet long and consisted of four rows, and the two center rows were used for evaluation. The previous crop was corn. Standard practices for grain corn production in Indiana were followed. Corn hybrid W2585SSRIB was planted in 30-inch row spacing at a rate of 34,000 seeds/acre on May 27. The field was overhead irrigated at 1 inch on August 5 and August 20. All fungicide applications were applied at silk (R1) growth stage on August 6 at 15 gal/acre and 40 psi using a Lee self-propelled sprayer equipped with a 10-foot boom, fitted with six TJ-VS 8002 nozzles spaced 20 inches apart. Disease ratings were assessed on September 14 and September 29 at dent (R5) and maturity (R6) growth stages, respectively. Tar spot was visually assessed as a percentage of stroma, as were percentage of symptomatic tissues (chlorosis and necrosis) per leaf on five plants in each plot at the ear leaf (EL), ear leaf minus two (EL−2), and ear leaf plus two (EL+2). Values for each plot were averaged before analysis. The two center rows of each plot were harvested on November 3, and yields were adjusted to 15.5% moisture. All data were analyzed in SAS 9.4 (SAS Institute, Cary, NC). A generalized linear mixed model analysis of variance was performed using PROC GLIMMIX. Values are least squares means, and values with different letters are significantly different based on least squares means test (α=0.05).

In 2021, weather conditions were favorable for disease. Tar spot was the most prominent disease in the trial and reached high severity. At R5 growth stage on September 14, tar spot stroma severity was significantly reduced on all leaves over the nontreated control by Delaro Complete, Delaro SC, Tilt and Veltyma, and Revytek on EL (Table 26). The percent of symptomatic tissue was significantly reduced on all leaves by Delaro Compete, Delaro SC, Headline Amp, and Veltyma. No significant differences were observed among fungicide treatments and the nontreated control for all disease ratings at R6 growth stage on September 29 (Table 27). All fungicides significantly increased percent of canopy green over the nontreated control except for Miravis Neo on September 14, but no significant differences among fungicide treatments and the nontreated control were observed on September 29 for percent of canopy green, lodging, moisture, test weight, and yield of corn (Table 28).

TABLE 26. *Effect of Fungicide on Tar Spot at Dent (R5) Growth Stage*

TREATMENT AND RATE/ACRE[z]	TAR SPOT[y] % EL-2	TAR SPOT[y] % EL	TAR SPOT[y] % EL+2	TAR SPOT CHLOR/ NEC[x] % EL-2	TAR SPOT CHLOR/ NEC[x] % EL	TAR SPOT CHLOR/ NEC[x] % EL+2
Nontreated control	30.0 a	22.5 a	20.8 a	65.8 ab	34.8 a	15.0 a
Revytek 3.33 LC at 8.0 fl oz	22.3 abc	14.5 cde	15.6 a-d	42.5 a-e	20.5 abc	6.0 bcd
Veltyma 3.24 S 7.0 fl oz	19.3 bc	14.0 cde	14.3 cde	31.5 cde	15.3 bcd	6.5 bcd
Headline 2.09 SC at 6.0 fl oz	25.5 ab	20.0 abc	17.0 abc	57.0 abc	28.3 abc	8.0 a-d
Headline AMP 1.68 SC at 10.0 fl oz	20.5 bc	14.3 cde	13.3 cde	40.3 b-e	15.8 bcd	4.8 bcd
Aproach Prima 2.34 SC at 6.8 fl oz	25.8 ab	17.8 abc	16.5 a-d	50.3 abc	23.5 abc	7.0 bcd
Miravis Neo 2.5 SE at13.7 fl oz	30.3 a	20.8 ab	18.0 abc	68.3 a	29.3 ab	12.3 ab
Delaro Complete 3.83 SC at 8.0 fl oz	16.8 cd	10.5 de	11.6 de	21.8 de	8.8 cd	3.5 cd
Delaro 325 SC at 8.0 fl oz	10.5 d	8.9 e	9.5 e	18.0 e	3.3 d	0.5 d
Lucento 4.17 SC at 5.0 fl oz	23.8 abc	19.8 abc	19.5 ab	46.0 a-d	25.5 ab	8.8 abc
Tilt 3.6 EC at 4.0 fl oz	19.0 bc	15.5 bcd	15.3 bcd	47.5 a-d	22.5 abc	6.0 bcd
P-value[w]	*0.0016*	*0.0010*	*0.0046*	*0.0107*	*0.0078*	*0.0388*

[z] Fungicides were applied at silk (R1) growth stage on August 6. All treatments applied contained a nonionic surfactant (Preference) at a rate of 0.25% v/v.

[y] Tar spot stroma was visually assessed as a percentage (0–100%) of leaf area on five plants in each plot at the ear leaf (EL), ear leaf minus two (EL-2), and ear leaf plus two (EL+2) on September 14.

[x] Tar spot chlorotic and necrotic symptoms were visually assessed as a percentage (0–100%) of leaf area on five plants in each plot at the ear leaf (EL), ear leaf minus two (EL-2), and ear leaf plus two (EL+2) on September 14.

[w] All data were analyzed in SAS 9.4 (SAS Institute, Cary, NC). A generalized linear mixed model analysis of variance was performed using PROC GLIMMIX. Values are least squares means, and values with different letters are significantly different based on least squares means test (α=0.05).

TABLE 27. *Effect of Fungicide on Tar Spot at Maturity (R6) Growth Stage*

TREATMENT AND RATE/ACRE[z]	TAR SPOT[y] % EL-2	TAR SPOT[y] % EL	TAR SPOT[y] % EL+2	TAR SPOT CHLOR/ NEC[x] % EL-2	TAR SPOT CHLOR/ NEC[x] % EL	TAR SPOT CHLOR/ NEC[x] % EL+2
Nontreated control	38.5	32.0	28.5	100.0	100.0	100.0
Revytek 3.33LC at 8.0 fl oz	32.8	28.0	24.5	100.0	100.0	98.8
Veltyma 3.24S 7.0 fl oz	32.8	28.3	26.3	100.0	96.0	93.3
Headline 2.09SC at 6.0 fl oz	35.3	27.8	26.5	100.0	100.0	99.0
Headline AMP 1.68SC at 10.0 fl oz	30.0	25.8	23.3	100.0	100.0	96.5
Aproach Prima 2.34SC at 6.8 fl oz	31.0	25.8	25.8	100.0	100.0	99.3
Miravis Neo 2.5SE at 13.7 fl oz	35.8	31.0	26.3	100.0	100.0	100.0
Delaro Complete 3.83SC at 8.0 fl oz	33.3	27.5	24.8	100.0	100.0	97.0
Delaro 325SC at 8.0 fl oz	35.8	29.5	26.8	100.0	100.0	97.4
Lucento 4.17SC at 5.0 fl oz	34.0	28.3	22.3	100.0	100.0	100.0
Tilt 3.6EC at 4.0 fl oz	34.5	31.3	27.5	100.0	100.0	98.0
P-value[w]	0.1497	0.2591	0.5059	—	0.4654	0.2378

[z] Fungicides were applied at R1 (silk) growth stage on August 6. All treatments applied contained a nonionic surfactant (Preference) at a rate of 0.25% v/v.

[y] Tar spot stroma was visually assessed as a percentage (0–100%) of leaf area on five plants in each plot at the ear leaf (EL), ear leaf minus two (EL-2), ear leaf plus two (EL+2) on September 29.

[x] Tar spot chlorotic and necrotic symptoms were visually assessed as a percentage (0–100%) of leaf area on five plants in each plot at the ear leaf (EL), ear leaf minus two (EL-2), ear leaf plus two (EL+2) on September 29.

[w] All data were analyzed in SAS 9.4 (SAS Institute, Cary, NC). A generalized linear mixed model analysis of variance was performed using PROC GLIMMIX. Values are least squares means, and values with different letters are significantly different based on least squares means test (α=0.05).

TABLE 28. *Effect of Fungicide on Canopy Greenness, Lodging, and Yield of Corn*

TREATMENT AND RATE/ACRE[z]	CANOPY%[y] 14 SEP	CANOPY%[y] 29 SEP	LODGING %[x] 29 SEP	HARVEST MOISTURE %	TEST WEIGHT LB/BU	YIELD[w] BU/ACRE
Nontreated control	63.8 d	0.0	32.5	19.6	52.9	135.6
Revytek 3.33LC at 8.0 fl oz	83.8 ab	0.5	15.0	20.9	52.3	139.2
Veltyma 3.24S 7.0 fl oz	82.5 ab	5.0	12.5	19.9	67.7	155.8
Headline 2.09SC at 6.0 fl oz	77.5 bc	1.3	12.5	20.8	53.4	149.0
Headline AMP 1.68SC at 10.0 fl oz	81.3 ab	2.5	5.0	20.9	52.2	145.6
Aproach Prima 2.34SC at 6.8 fl oz	78.8 bc	1.0	7.5	21.2	51.9	145.8
Miravis Neo 2.5SE at 13.7 fl oz	68.8 cd	0.0	15.0	19.8	53.6	146.3
Delaro Complete 3.83SC at 8.0 fl oz	87.5 ab	3.0	2.5	22.0	51.8	154.3
Delaro 325SC at 8.0 fl oz	91.3 a	2.3	10.0	21.7	52.1	141.9
Lucento 4.17SC at 5.0 fl oz	77.5 bc	0.0	7.5	20.4	51.8	138.5
Tilt 3.6EC at 4.0 fl oz	81.3 ab	1.8	5.0	20.9	52.2	147.8
P-value[v]	*0.0039*	*0.1470*	*0.4175*	*0.1729*	*0.4386*	*0.0973*

[z] Fungicides were applied at silk (R1) growth stage on August 6. All treatments applied contained a nonionic surfactant (Preference) at a rate of 0.25% v/v.

[y] Canopy greenness was visually assessed as a percentage (0–100%) of crop canopy green.

[x] Lodging was visually assessed as a percentage of lodged stalks when pushed from shoulder height to 45° from vertical.

[w] Yields were adjusted to 15.5% moisture after harvesting on November 3.

[v] All data were analyzed in SAS 9.4 (SAS Institute, Cary, NC). A generalized linear mixed model analysis of variance was performed using PROC GLIMMIX. Values are least squares means, and values with different letters are significantly different based on least squares means test (α=0.05).

EVALUATION OF FUNGICIDE TIMING FOR TAR SPOT MANAGEMENT IN CORN IN NORTHWESTERN INDIANA, 2021 (COR21-03.PPAC)

T. J. Ross, S. Shim, and D. E. P. Telenko, Department of Botany and Plant Pathology, Purdue University West Lafayette, IN 47907-2054

CORN (*ZEA MAYS* W2585SSRIB)

Tar spot, *Phyllachora maydis*

A trial was established at the Pinney Purdue Agricultural Center (PPAC) in Porter County, Indiana. The experiment was a randomized complete block design with four replications. Plots were 10 feet wide and 30 feet long and consisted of four rows, and the two center rows were used for evaluation. The previous crop was corn. Standard practices for grain corn production in Indiana were followed. Corn hybrid W2585SS-RIB was planted in 30-inch row spacing at a rate of 34,000 seeds/acre on May 27. The field was overhead irrigated at 1 inch on August 5 and August 20. All fungicide applications were applied at 15 gal/acre and 40 psi using a Lee self-propelled sprayer equipped with a 10-foot boom, fitted with six TJ-VS 8002 nozzles spaced 20 inches apart. Fungicides were applied on July 23, August 2, August 6, August 20, August 30, September 10, and September 16 at eight-leaf (V8), 12-leaf (V12), silk (VT/R1), blister (R2), milk (R3), dough (R4), and dent (R5), and V8 followed by VT (V8 followed by VT) growth stages, respectively. A weather-based prediction model for tar spot was used to predict fungicide timing, which triggered a 12-leaf (V12) application on August 2. Disease ratings were assessed on September 14 and September 29 at dent (R5) and maturity (R6) growth stages, respectively. Tar spot was rated by visually assessing the percentage of stroma and the percentage of symptomatic tissues (chlorosis and necrosis) per leaf on five plants in each plot at the ear leaf (EL), ear leaf minus two (EL−2), and ear leaf plus two (EL+2). Values for each plot were averaged before analysis. The two center rows of each plot were harvested on November 3, and yields were adjusted to 15.5% moisture. All data were analyzed in SAS 9.4 (SAS Institute, Cary, NC). A generalized linear mixed model analysis of variance was performed using PROC GLIMMIX. Values are least squares means, and values with different letters are significantly different based on least squares means test (α=0.05).

In 2021, weather conditions were favorable for disease. Tar spot was the most prominent disease in the trial and reached high severity. Trivapro applied at R2 and R3 growth stages and with Tarspotter at V12 significantly reduced tar spot stroma on all leaves over the nontreated control except for Tarspotter V12 on EL at R5 (Table 29). Trivapro was applied at R2 and R3 and at Tarspotter (V2). All treatments significantly reduced symptomatic tissue severity on the EL−2, EL, and EL+2 over the nontreated control on September 14. In addition, application at V8 followed by R1 reduced tar spot symptoms on EL and EL+2 on September 14. No significant differences were observed among fungicide treatments and the nontreated control for stroma severity on the EL−2, but all application timings significantly reduced stroma severity over the nontreated control on the EL and EL−2 at R6 on September 29 (Table 30). On September 29, R3, R4 and Tarspotter timed applications significantly reduced symptomatic tissue severity on the EL−2, whereas V12, R2, R3, R4, and Tarspotter significantly reduced symptomatic tissue severity on the EL, and V12, R2, R3, R4, and R5 applications significantly reduced symptomatic tissue severity on the EL+2. All application timings significantly increased percent canopy green over the nontreated control except for applications made at the V12, VT/R1, and R5 growth stages on September 24. On September 29 all timings except V8, VT/R1, and V8 followed by R1 significantly increased percent of canopy green over the nontreated control (Table 31). No significant differences were detected for moisture and test weight. Trivapro applied at the VT/R1, R2, R3, and R4 growth stages significantly increase yield of corn over the nontreated control.

TABLE 29. *Effect of Fungicide on Tar Spot at Dent (R5) Growth Stage*

TREATMENT AND RATE/ACRE[z]	TAR SPOT[y] % EL−2	TAR SPOT[y] % EL	TAR SPOT[y] % EL+2	TAR SPOT CHLOR/ NEC[x] % EL−2	TAR SPOT CHLOR/ NEC[x] % EL	TAR SPOT CHLOR/ NEC[x] % EL+2
Nontreated control	14.5 bc	12.2 abc	9.9 ab	15.3 bc	4.8 bcd	0.5 bc
Trivapro 2.21 SE 13.7 fl oz at V8	10.9 cd	10.2 cd	7.9 bc	5.5 cd	2.3 cde	0.5 bc
Trivapro 2.21 SE 13.7 fl oz at V12	9.4 cd	9.0 cde	7.6 bc	6.0 cd	1.5 de	0.0 c
Trivapro 2.21 SE 13.7 fl oz at VT/R1	17.8 ab	14.5 ab	11.8 a	14.3 bc	6.0 abc	3.3 a
Trivapro 2.21 SE 13.7 fl oz at R2	8.3 d	5.7 e	4.2 d	3.5 d	0.3 e	0.0 c
Trivapro 2.21 SE 13.7 fl oz at R3	7.7 d	6.6 de	4.3 d	1.0 d	0.0 e	0.0 c
Trivapro 2.21 SE 13.7 fl oz at R4	18.0 ab	14.6 ab	9.9 ab	19.0 b	7.3 ab	2.0 ab
Trivapro 2.21 SE 13.7 fl oz at R5	20.8 a	15.5 a	11.0 a	29.8 a	9.8 a	2.5 a
Trivapro 2.21 SE 13.7 fl oz at V8 fb R1	11.0 cd	11.0 bc	7.9 bc	4.3 d	2.0 cde	0.0 c
Trivapro 2.21 SE 13.7 fl oz at Tar spot model (V12)	7.1 d	9.0 cde	6.4 cd	2.8 d	0.3 e	0.0 c
P-value[w]	*0.0001*	*0.0001*	*0.0001*	*0.0001*	*0.0002*	*0.0016*

[z] Fungicide treatments were applied on July 23, August 2, August 6, August 20, August 30, September 10, and September 16 at eight-leaf (V8), 12-leaf (V12), silk (VT/R1), blister (R2), milk (R3), dough (R4), and dent (R5), and V8 followed by VT (V8 fb VT) growth stages, respectively. Tar spot model = tar spot weather-based model application. The tar spot model triggered application at V12 growth stage on August 2, and only a single application was made. All treatments contained a nonionic surfactant (Preference) at a rate of 0.25% v/v. fb = followed by.

[y] Tar spot stroma was visually assessed as a percentage (0–100%) of leaf area on five plants in each plot at the ear leaf (EL), ear leaf minus two (EL−2), ear leaf plus two (EL+2) on September 14.

[x] Tar spot chlorotic and necrotic symptoms were visually assessed as a percentage (0–100%) of leaf area on five plants in each plot at the ear leaf (EL), ear leaf minus two (EL−2), and ear leaf plus two (EL+2) on September 14.

[w] All data were analyzed in SAS 9.4 (SAS Institute, Cary, NC). A generalized linear mixed model analysis of variance was performed using PROC GLIMMIX. Values are least squares means, and values with different letters are significantly different based on least squares means test (α=0.05).

TABLE 30. *Effect of Fungicide on Tar Spot at Maturity (R6) Growth Stage*

TREATMENT AND RATE/ACRE[z]	TAR SPOT[y] % EL−2	TAR SPOT[y] % EL	TAR SPOT[y] % EL+2	TAR SPOT CHLOR/ NEC[x] % EL−2	TAR SPOT CHLOR/ NEC[x] % EL	TAR SPOT CHLOR/ NEC[x] % EL+2
Nontreated control	28.3	29.0 a	24.0 a	100.0 a	98.0 a	78.8 a
Trivapro 2.21 SE 13.7 fl oz at V8	26.0	22.8 b	17.8 b	94.8 ab	86.8 ab	66.8 ab
Trivapro 2.21 SE 13.7 fl oz at V12	30.3	20.3 bc	15.8 bc	90.0 ab	70.8 bc	51.8 bc
Trivapro 2.21 SE 13.7 fl oz at VT/R1	24.0	22.0 b	19.3 b	100.0 a	91.0 ab	66.5 ab
Trivapro 2.21 SE 13.7 fl oz at R2	21.3	15.5 cd	12.3 cd	79.0 ab	39.3 de	3.0 e
Trivapro 2.21 SE 13.7 fl oz at R3	15.9	8.7 e	7.4 e	24.3 d	1.0 f	0.3 e
Trivapro 2.21 SE 13.7 fl oz at R4	24.5	14.8 d	10.5 de	76.3 b	52.5 cd	13.8 de
Trivapro 2.21 SE 13.7 fl oz at R5	21.3	22.0 b	16.8 b	96.3 ab	70.0 bc	34.3 cd
Trivapro 2.21 SE 13.7 fl oz at V8 fb R1	23.5	21.3 b	19.0 b	100.0 a	88.3 ab	64.8 ab
Trivapro 2.21 SE 13.7 fl oz at Tarspotter (V12)	21.3	13.0 de	8.9 de	53.3 c	18.3 ef	2.8 e
P-value[w]	0.3507	0.0001	0.0001	0.0001	0.0001	0.0001

[z] Fungicide treatments were applied on July 23, August 2, August 6, August 20, August 30, September 10, and September 16 at eight-leaf (V8), 12-leaf (V12), silk (VT/R1), blister (R2), milk (R3), dough (R4), dent (R5), V8 followed by VT (V8 fb VT) growth stages, respectively. Tarspotter = tar spot weather-based model application. The tar spot model triggered application at V12 growth stage. All treatments contained a nonionic surfactant (Preference) at a rate of 0.25% v/v. fb = followed by.

[y] Tar spot stroma was visually assessed as a percentage (0–100%) of leaf area on five plants in each plot at the ear leaf (EL), ear leaf minus two (EL−2), and ear leaf plus two (EL+2) on September 29.

[x] Tar spot chlorotic and necrotic symptoms were visually assessed as a percentage (0–100%) of leaf area on five plants in each plot at the ear leaf (EL), ear leaf minus two (EL−2), and ear leaf plus two (EL+2) on September 29.

[w] All data were analyzed in SAS 9.4 (SAS Institute, Cary, NC). A generalized linear mixed model analysis of variance was performed using PROC GLIMMIX. Values are least squares means, and values with different letters are significantly different based on least squares means test (α=0.05).

TABLE 31. *Effect of Fungicide on Canopy Greenness and Yield of Corn*

TREATMENT, RATE/ACRE, AND TIMING[z]	CANOPY[y] %	CANOPY[y] %	HARVEST MOISTURE %	TEST WEIGHT LB/BU	YIELD[x] BU/ACRE
Nontreated control	55.0 d	36.3 d	22.1	53.5	97.2 b
Trivapro 2.21 SE 13.7 fl oz at V8	61.3 cd	40.0 d	21.7	53.0	112.1 ab
Trivapro 2.21 SE 13.7 fl oz at V12	67.5 bc	62.5 c	22.6	52.9	120.0 ab
Trivapro 2.21 SE 13.7 fl oz at VT/R1	63.8 cd	40.0 d	21.8	53.2	125.2 a
Trivapro 2.21 SE 13.7 fl oz at R2	93.8 a	86.3 ab	22.8	62.7	128.3 a
Trivapro 2.21 SE 13.7 fl oz at R3	96.5 a	95.0 a	24.4	53.1	130.1 a
Trivapro 2.21 SE 13.7 fl oz at R4	76.3 b	75.0 abc	21.8	53.6	135.9 a
Trivapro 2.21 SE 13.7 fl oz at R5	65.0 cd	67.5 bc	22.2	53.3	115.5 ab
Trivapro 2.21 SE 13.7 fl oz at V8 fb R1	66.3 bc	33.8 d	22.6	52.9	116.2 ab
Trivapro 2.21 SE 13.7 fl oz at Tarspotter (V12)	92.5 a	88.8 a	22.6	53.3	122.6 ab
P-value[w]	0.0001	0.0001	0.0886	0.4448	0.1877

[z] Fungicide treatments were applied on July 23, August 2, August 6, August 20, August 30, September 10, and September 16 at eight-leaf (V8), 12-leaf (V12), silk (VT/R1), blister (R2), milk (R3), dough (R4), dent (R5), and V8 followed by VT (V8 fb VT) growth stages, respectively. Tarspotter = tar spot weather-based model application. The tar spot model triggered application at V12. All treatments contained a nonionic surfactant (Preference) at a rate of 0.25% v/v. fb = followed by.

[y] Canopy greenness was visually assessed as a percentage (0–100%) of crop canopy green on September 24.

[x] Yields were adjusted to 15.5% moisture after harvesting on November 3.

[w] All data were analyzed in SAS 9.4 (SAS Institute, Cary, NC). A generalized linear mixed model analysis of variance was performed using PROC GLIMMIX. Values are least squares means, and values with different letters are significantly different based on least squares means test (α=0.05).

FUNGICIDE EVALUATION FOR TAR SPOT IN ORGANIC CORN IN NORTHWESTERN INDIANA, 2021 (COR21-05.PPAC)

C. R. Da Silva, S. Shim, and D. E. P. Telenko, Department of Botany and Plant Pathology, Purdue University West Lafayette, IN 47907-2054

CORN (*ZEA MAYS* ALSEED O.84–95UP)

Tar spot, *Phyllachora maydis*

A trial was established at the Pinney Purdue Agricultural Center (PPAC) in Porter County, Indiana. The experiment was a randomized complete block design with four replications. Plots were 10 feet wide and 30 feet long and consisted of four rows, and the two center rows were used for evaluation. The previous crop was corn. Standard practices for organic grain corn production in Indiana were followed. Corn organic hybrid ALSEED O.84–95UP was planted in 30-inch row spacing at a rate of 34,000 seeds/acre on May 25. The field was overhead irrigated weekly at 1 inch unless weekly rainfall was 1 inch or higher to encourage disease. All fungicide applications were applied at 15 gal/acre and 40 psi using a Lee self-propelled sprayer equipped with a 10-foot boom, fitted with six TJ-VS 8002 nozzles spaced 20 inches apart. Fungicide treatments were applied on August 2 at silk (R1) growth stage. Disease ratings were assessed on September 16 at dent (R5) growth stage. Tar spot was rated by visually assessing the percentage of stroma (0–100%) and the percentage of symptomatic tissues (chlorosis and necrosis) (0–100%) per leaf on five plants in each plot at the ear leaf. Northern corn leaf blight (NCLB) was rated for disease severity by visually assessing the percentage of symptomatic leaf area in the mid canopy. Values for the five leaves were averaged before analysis. Percent canopy green was rated by visually assessing the percentage (0–100%) of the whole plot for crop canopy that remained green at dent (R5) growth stage. The two center rows of each plot were harvested on November 3, and yields were adjusted to 15.5% moisture. All data were analyzed in SAS 9.4 (SAS Institute, Cary, NC). A generalized linear mixed model analysis of variance was performed using PROC GLIMMIX. Values are least squares means, and values with different letters are significantly different based on least squares means test (α=0.05).

In 2021, weather conditions were favorable for disease. Tar spot was the most prominent disease in the trial and reached high severity. All fungicide treatments reduced tar spot stroma severity on ear leaf over the non-treated control (Table 32). Headline Amp significantly reduced the percentage of symptomatic tissues on ear leaf. There was no significant difference between treatments for severity of NCLB on ear leaf. Headline Amp had the highest percent of green plots and corn yield. There were no significant differences between treatments for harvest moisture and test weight.

TABLE 32. *Effect of Fungicide on Foliar Disease Severity at Dent (R5) Growth Stage, Canopy Greenness, and Yield of Corn*

TREATMENT AND RATE/ACRE[z]	TAR SPOT[y] %	TAR SPOT CHLOR/ NEC[x] %	NCLB[w] %	CANOPY[v] %	HARVEST MOISTURE %	TEST WEIGHT LB/BU	YIELD[u] BU/ACRE
Nontreated control	25.8 a	86.0 a	0.5 a	23.8 b	16.8	55.9	148.2 b
Headline AMP 1.68 SE 10.0 fl oz	10.0 d	56.3 b	1.2 a	40.0 a	17.0	56.4	162.6 a
Serifel WP 16.0 fl oz	18.3 bc	86.8 a	0.5 a	22.5 b	16.8	56.2	148.3 b
Actinovate AG 12.0 ox	20.3 b	76.3 ab	0.0 a	25.0 b	16.7	55.6	160.7 ab
Badge X2 SC 1.8 lb	14.8 cd	74.3 ab	1.9 a	33.8 b	17.1	56.2	149.9 b
OxiDate 5.0 128.0 fl oz	20.3 b	78.3 ab	0.5 a	23.8 b	16.9	56.3	159.4 ab
P-value[t]	0.0001	0.0001	0.6753	0.0001	0.4366	0.2614	0.0001

[z] Fungicide treatments were applied at on August 2 at R1 (silk) growth stage.

[y] Tar spot stroma was visually assessed as a percentage (0–100%) of leaf area on five plants in each plot at the ear leaf on September 16.

[x] Tar spot chlorotic and necrotic symptoms were visually assessed as a percentage (0–100%) of leaf area on five plants in each plot at the ear leaf on September 16.

[w] NCLB was visually assessed as a percentage of symptomatic leaf area in the midcanopy on September 16. NCLB = northern corn leaf blight.

[v] Canopy greenness was visually assessed as a percentage (0–100%) of crop canopy green on September 16.

[u] Yields were adjusted to 15.5% moisture after harvesting on November 3.

[t] All data were analyzed in SAS 9.4 (SAS Institute, Cary, NC). A generalized linear mixed model analysis of variance was performed using PROC GLIMMIX. Values are least squares means, and values with different letters are significantly different based on least squares means test (α=0.05).

FUNGICIDE TIMING AND APPLICATION FOR TAR SPOT IN CORN IN NORTHWESTERN INDIANA, 2021 (COR21-06.PPAC)

C. R. Da Silva, S. Shim, and D. E. P. Telenko, Department of Botany and Plant Pathology, Purdue University West Lafayette, IN 47907-2054

CORN (*ZEA MAYS* W2585SSRIB)

Tar spot, *Phyllachora maydis*

A trial was established at the Pinney Purdue Agricultural Center (PPAC) in Porter County, Indiana. The experiment was a randomized complete block design with four replications. Plots were 10 feet wide and 30 feet long and consisted of four rows, and the two center rows were used for evaluation. The previous crop was corn. Standard practices for grain corn production in Indiana were followed. Corn hybrid W2585SSRIB was planted in 30-inch row spacing at a rate of 34,000 seeds/acre on May 27. The field was overhead irrigated weekly at 1 inch unless weekly rainfall was 1 inch or higher to encourage disease. All fungicide applications were applied at 15 gal/acre and 40 psi. Fungicides were applied at first detection of tar spot and at V8, tassel/silk (VT/R1), and milk (R3) growth stages on July 14, July 23, August 5, and August 30, respectively. To compare a single versus a two-pass fungicide application programs, at three week after treatment (3-WAT) fungicide was applied following the initial application. The 3-WAT applications occurred on August 2, August 12, August 27, and September 16. Disease ratings were assessed on September 24 at dent (R5) growth stage. Tar spot was visually assessed as a percentage of stroma (0–100%) and a percentage of symptomatic tissues (chlorosis and necrosis) (0–100%) per leaf on five plants in each plot at the ear leaf. Values for the five leaves were averaged before analysis. Percent canopy green was rated by visually assessing the percentage (0–100%) of the whole plot for crop canopy that remained green at dent (R5) growth stage. The two center rows of each plot were harvested on November 3, and yields were adjusted to 15.5% moisture. All data were analyzed in SAS 9.4 (SAS Institute, Cary, NC). A generalized linear mixed model analysis of variance was performed using PROC GLIMMIX. Values are least squares means, and values with different letters are significantly different based on least squares means test (α=0.05).

In 2021 weather conditions were favorable for tar spot, which reached high severity. Veltyma applied at V8, first detection followed by (fb) 3-WAT, V8 fb 3-WAT, VT fb 3-WAT, and R3 fb 3-WAT significantly reduced tar spot stroma severity over the nontreated controls on the ear leaf (Table 33). In addition, Lucento applied at V8 fb 3-WAT and at VT fb 3-WAT reduced tar spot. A single application of Veltyma at V8 and R3 reduced chlorotic and necrotic symptoms over nontreated control; in addition, tar spot symptoms were reduced by Veltyma at the first detection of tar spot fb 3-WAT, V8 fb 3-WAT, VT fb 3-WAT, and R3 fb 3-WAT and Lucento at R3, V8 fb 3-WAT, VT fb 3-WAT, and R3 fb 3-WAT. All fungicide programs of Veltyma and Lucento significantly increased the percentage of green canopy over the nontreated control. Veltyma when applied at R3, V8 fb 3-WAT, VT fb 3-WAT, R3 fb 3-WAT and Lucento when applied at V8 fb 3-WAT, VT fb 3-WAT, and R3 fb 3-WAT increased yield over the nontreated control.

TABLE 33. *Effect of Fungicide on Tar Spot of Corn at Dent (R5) Growth Stage, Canopy Greenness, and Yield of Corn*

TREATMENT, RATE/ACRE, AND TIMING[z]	TAR SPOT[y] %	TAR SPOT CHLOR/NEC[x] %	CANOPY[w] %	HARVEST MOISTURE %	TEST WEIGHT LB/BU	YIELD[v] BU/ACRE
Nontreated control	31.3 a	90.5 a	30.0 i	21.15	52.85	143.2 d
Veltyma 3.34 S 7.0 fl oz at first detection	28.5 ab	75.0 a	40.0 h	20.40	53.38	149.3 cd
Veltyma 3.34 S 7.0 fl oz at V8	17.4 efg	45.2 b	60.0 def	23.15	52.98	157.6 bcd
Veltyma 3.34 S 7.0 fl oz at VT	26.3 a-d	75.5 a	42.5 h	21.99	52.15	157.1 bcd
Veltyma 3.34 S 7.0 fl oz at R3	22.3 b-e	32.5 b-e	62.5 bcd	21.98	53.03	169.8 ab
Veltyma 3.34 S 7.0 fl oz at first detection fb 3-WAT	18.3 efg	41.7 bc	61.3 cde	21.75	53.23	149.4 cd
Veltyma 3.34 S 7.0 fl oz at V8 fb 3-WAT	0.5 h	0.0 f	91.3 a	23.95	52.63	184.2 a
Veltyma 3.34 S 7.0 fl oz at VT fb 3-WAT	11.9 g	12.8 ef	68.3 bc	22.03	53.40	188.2 a
Veltyma 3.34 S 7.0 fl oz at R3 fb 3-WAT	20.6 def	30.2 b-e	52.5 fg	21.85	53.30	176.8 ab
Nontreated control	27.8 abc	86.5 a	31.3 i	22.00	53.60	144.5 d
Lucento 7.17 SC 5.0 fl oz at first detection	27.5 abc	78.5 a	42.5 h	21.70	53.50	149.1 cd
Lucento 7.17 SC 5.0 fl oz at V8	25.5 a-d	77.2 a	40.0 h	16.26	52.68	160.9 bcd
Lucento 7.17 SC 5.0 fl oz at VT	26.2 a-d	73.2 a	41.3 h	21.23	52.05	150.9 cd
Lucento 7.17 SC 5.0 fl oz at R3	21.0 cde	14.6 def	65.0 bcd	21.35	52.68	159.5 bcd
Lucento 7.17 SC 5.0 fl oz at first detection fb 3-WAT	26.5 a-d	70.5 a	46.3 gh	21.58	53.58	158.0 bcd
Lucento 7.17 SC 5.0 fl oz at V8 fb 3-WAT	14.1 fg	25.1 cde	70.0 b	21.75	52.98	175.0 ab
Lucento 7.17 SC 5.0 fl oz at VT fb 3-WAT	16.1 efg	33.0.bcd	57.5 def	21.85	53.70	163.4 bc
Lucento 7.17 SC 5.0 fl oz at R3 fb 3-WAT	25.5 a-d	34.0 bcd	53.8 efg	21.45	53.48	164.2 bc
P-value[u]	*0.0001*	*0.0001*	*0.0001*	*0.3341*	*0.3289*	*0.0001*

[z] Fungicides were applied at first detection of tar spot and at V8, tassel/silk (VT/R1), and milk (R3) growth stages on July 14, July 23, August 5, and August 30, respectively. The second application occurred 3 weeks after treatment (3-WAT) on August 2, August 12, August 27, and September 16. All treatments contained a nonionic surfactant (Preference) at VT or later applications at a rate of 0.25% v/v. fb = followed by, 3-WAT = three weeks after treatment. .

[y] Tar spot stroma was visually assessed as percentage (0–100%) of leaf area on five plants in each plot at the ear leaf on September 24.

[x] Tar spot chlorotic and necrotic symptoms were visually assessed as a percentage (0–100%) of leaf area on five plants in each plot at the ear leaf on September 24.

[w] Canopy greenness was visually assessed as a percentage (0–100%) of crop canopy green on September 24.

[v] Yields were adjusted to 15.5% moisture after harvesting on November 3.

[u] All data were analyzed in SAS 9.4 (SAS Institute, Cary, NC). A generalized linear mixed model analysis of variance was performed using PROC GLIMMIX. Values are least squares means, and values with different letters are significantly different based on least squares means test (α=0.05).

EVALUATION OF EFFECT OF TILLAGE AND CULTIVAR FOR FOLIAR DISEASES RISK IN CORN, 2021 (COR21-08.PPAC)

S. Shim and D. E. P. Telenko, Department of Botany and Plant Pathology, Purdue University West Lafayette, IN 47907-2054

CORN (*ZEA MAYS* W2585SSRIB, P0589AMXT)

Tar spot, *Phyllachora maydis*

A trial was established at the Pinney Purdue Agricultural Center (PPAC) in Porter County, Indiana. The experiment was a split-plot with four replications. Plots were 7.5 feet wide and 30 feet long and consisted of six rows, and the two center rows were used for evaluation. The previous crop was corn. Standard practices for grain corn production in Indiana were followed. Corn hybrids W2585SSRIB and P0589AMXT were planted in 30-inch row spacing at a rate of 32,000 seeds/acre on May 27. Standard practices for nonirrigated grain corn production in Indiana were followed. No foliar fungicides were applied. Disease ratings were assessed on September 14 and September 28 at dent (R5) and maturity (R6) growth stages, respectively. Tar spot was visually rated as a percentage of stroma and a percentage of symptomatic tissues (chlorosis and necrosis) per leaf on 20 plants in each plot at the ear leaf minus two (EL−2), ear leaf minus one (EL−1), ear leaf (EL), ear leaf plus one (EL+1). Values for each plot were averaged before analysis. The four center rows of each plot were harvested on November 4, and yields were adjusted to 15.5% moisture All data were analyzed in SAS 9.4 (SAS Institute, Cary, NC). A generalized linear mixed model analysis of variance was performed using PROC GLIMMIX. Values are least squares means, and values with different letters are significantly different based on least squares means test (α=0.05).

In 2021, weather conditions were favorable for disease. Tar spot was the most prominent disease in the trial and reached high severity. Tar spot stroma severity and chlorotic and necrotic symptoms were significantly reduced with tar spot moderate resistant hybrid (P0589AMXT) compared to tar spot susceptible hybrid (W2585SSRIB) on all leaves on September 14 (Table 34). Tar spot stroma severity was significantly reduced with tillage treatment (low residue) compared to no-tillage (high residue) on the EL−1 on September 14. Tar spot chlorotic and necrotic symptoms were also reduced with tillage treatment compared to no-tillage treatment on the EL−2 and EL−1 on September 14. On September 28, tar spot stroma severity was significantly reduced with tar spot moderate resistant hybrid (P0589AMXT) compared to tar spot susceptible hybrid (W2585SSRIB) on all leaves on September 28 (Table 35). Tar spot stroma severity was significantly reduced with tillage treatment (low residue) compared to no-tillage (high residue) on the EL−1 on September 28. Percent canopy greenness was increased with tar spot resistant hybrid (P0589AMXT) compared to tar spot susceptible hybrid (W2585SSRIB) on September 28. Test weight was higher with tillage treatment (low residue), and there were no significant differences on effect of tillage and hybrid for harvest moisture and yield of corn.

TABLE 34. *Effect of Tillage and Hybrid for Foliar Disease Risk in Corn at Dent (R5) Growth Stage on September 14*

TREATMENT[z]	TAR SPOT %[y]				TAR SPOT % CHLOR/NEC[x]			
	EL-2	EL-1	EL	EL+1	EL-2	EL-1	EL	EL+1
Tillage								
No tillage (high residue)	16.2	13.2 a	10.1	5.0	25.6 a	17.3 a	10.0	3.8
Yes tillage (low residue)	11.2	8.9 b	7.3	6.4	14.0 b	9.1 b	6.3	3.9
Hybrid								
P0589AMXT	6.0 b	5.2 b	4.6 b	3.9 b	4.9 b	3.3 b	2.4 b	1.4 b
W2585SSRIB	21.4 a	17.0 a	12.8 a	7.4 a	34.7 a	23.0 a	13.9 a	6.3 a
P-value (*tillage*)[w]	0.0502	0.0445	0.0984	0.1698	0.0244	0.0223	0.1008	0.8559
P-value (*hybrid*)	0.0001	0.0001	0.0001	0.0001	0.0001	0.0001	0.0001	0.0001
P-value (*tillage*hybrid*)	0.0548	0.0703	0.1031	0.0093	0.0145	0.0341	0.1396	0.4102

[z] No foliar fungicides were applied.

[y] Tar spot stroma was visually assessed as a percentage (0–100%) of leaf area on 20 plants in each plot at the ear leaf minus two (EL-2), ear leaf minus one (EL-1), ear leaf (EL), and ear leaf plus one (EL+1) on September 14.

[x] Tar spot chlorotic and necrotic symptoms were visually assessed as a percentage (0–100%) of leaf area on 20 plants in each plot at the ear leaf minus two (EL-2), ear leaf minus one (EL-1), ear leaf (EL), and ear leaf plus one (EL+1) on September 14.

[w] All data were analyzed in SAS 9.4 (SAS Institute, Cary, NC). A generalized linear mixed model analysis of variance was performed using PROC GLIMMIX. Values are least squares means, and values with different letters are significantly different based on least squares means test (α=0.05).

TABLE 35. *Effect of Tillage and Hybrid for Tar Spot Severity, Canopy Greenness, and Yield of Corn*

	TAR SPOT %[y]				CANOPY[x] %	HARVEST MOISTURE	TEST WEIGHT	YIELD[w]
	EL-2	EL-1	EL	EL+1				
Tillage								
No-tillage (high residue)	19.6	19.8 a	18.8	17.4	23.3	19.3	54.1	265.6
Yes-tillage (low residue)	15.8	16.2 b	17.1	16.9	14.9	18.6	56.5	308.2
Hybrid								
P0589AMXT	12.4 b	12.9 b	13.0 b	11.5 b	30.0 a	18.8	56.2	286.9
W2585SSRIB	23.0 a	23.1 a	22.9 a	22.8 a	8.3 b	19.1	54.4	287.0
P-value (*tillage*)[v]	0.0573	0.0239	0.0537	0.6249	0.1032	0.2349	0.0441	0.0763
P-value (*hybrid*)[v]	0.0001	0.0001	0.0001	0.0001	0.0001	0.2919	0.0798	0.9972
P-value (*tillage*hybrid*)[v]	0.4366	0.3155	0.4702	0.1209	0.0717	0.7990	0.3307	0.9044

[z] No foliar fungicides were applied.

[y] Tar spot stroma was visually assessed as a percentage (0–100%) of leaf area on 20 plants in each plot at the ear leaf minus two (EL-2), ear leaf minus one (EL-1), ear leaf (EL), and ear leaf plus one (EL+1) on September 28.

[x] Canopy greenness was visually assessed as a percentage (0–100%) of crop canopy green at maturity (R6) growth stage on September 28.

[w] Yields were adjusted to 15.5% moisture after harvesting on November 4.

[v] All data were analyzed in SAS 9.4 (SAS Institute, Cary, NC). A generalized linear mixed model analysis of variance was performed using PROC GLIMMIX. Values are least squares means, and values with different letters are significantly different based on least squares means test (α=0.05).

FUNGICIDE COMPARISON FOR FOLIAR DISEASES IN CORN IN NORTHWESTERN INDIANA, 2021 (COR21-15.PPAC)

S. Shim and D. E. P. Telenko, Department of Botany and Plant Pathology, Purdue University West Lafayette, IN 47907-2054

CORN (*ZEA MAYS* W2585SSRIB)

Tar spot, *Phyllachora maydis*

A trial was established at the Pinney Purdue Agricultural Center (PPAC) in Porter County, Indiana. The experiment was a randomized complete block design with four replications. Plots were 10 feet wide and 30 feet long and consisted of four rows, and the two center rows were used for evaluation. The previous crop was corn. Standard practices for grain corn production in Indiana were followed. Corn hybrid W2585SSRIB was planted in 30-inch row spacing at a rate of 34,000 seeds/acre on May 27. The field was overhead irrigated at 1 inch on August 5 and August 20. All fungicide applications were applied at 15 gal/acre and 40 psi using a Lee self-propelled sprayer equipped with a 10-foot boom, fitted with six TJ-VS 8002 nozzles spaced 20 inches apart. Fungicides were applied at V12, silk (R1), blister (R2), and milk (R3) growth stages on August 2, August 6, August 20, and August 30, respectively. Disease ratings were assessed on September 22 and September 29 at dent (R5), and maturity (R6) growth stages, respectively. Tar spot was rated by visually assessing the percentage of stroma, and percentage of symptomatic tissues (chlorosis and necrosis) per leaf on five plants in each plot at the ear leaf. Values for each plot were averaged before analysis. The two center rows of each plot were harvested on November 3, and yields were adjusted to 15.5% moisture. All data were analyzed in SAS 9.4 (SAS Institute, Cary, NC). A generalized linear mixed model analysis of variance was performed using PROC GLIMMIX. Values are least squares means, and values with different letters are significantly different based on least squares means test (α=0.05).

In 2021, weather conditions were favorable for disease. Tar spot was the most prominent disease in the trial and reached high severity. On September 22, all fungicides reduced tar spot stroma severity over the nontreated control except Miravis Neo at R1, Trivapro at R1, Veltyma at R1, Zolera at R1, and Vacciplant at R1 and R2 (Table 36). Miravis Neo at V12 and V12 followed by (fb) R3, Zolera + Vacciplant at R2, Veltyma at R2, and Delaro Complete at R2 reduced chlorosis and necrosis over the nontreated controls at R5. On September 29, Miravis Neo applied at V12 and V12 fb R3, Zolera + Vacciplant at R1 and R2, Zolera at R2, Veltyma at R2, and Delaro Complete at R2 reduced tar spot stroma over the nontreated controls. In addition, Brixen at all rates and Zolera at R1 reduced stroma. Miravis Neo at V12 and V12 fb R3, Delaro Complete at R1 and R2, Brixen all rates at R1, Zolera at R1 and R2, Zolera + Vacciplant at R1 and R2, and Veltyma at R2 increased canopy greenness of corn on September 22 (R5). By September 22 (R6), only treatments of Miravis Neo at V12 fb R3, Zolera at R1 and R2, Zolera + Vacciplant at R2, Veltyma at R2, and Delaro Complete at R2 were significantly greener than nontreated controls. Corn yield was highest in plots treated with Veltyma at R2, Delaro Complete at R2, Miravis Neo at V12 fb R3, Miravis Neo at V12, Veltyma at R1, Zolera at R1 and R2, and Zolera + Vacciplant over the nontreated controls (Table 37).

TABLE 36. *Effect of Fungicide on Tar Spot, Canopy Greenness, and Yield of Corn*

TREATMENT, RATE/ACRE, AND TIMING[z]	TAR SPOT[y] % SEP 22	TAR SPOT CHLOR/ NEC[x] % SEP 22	TAR SPOT[y] % SEP 29	CANOPY[w] % SEP 22	CANOPY[w] % SEP 29
Nontreated control	30.8 a	92.2 a	25.0 a	45.0 fg	2.0 f
Miravis Neo 2.5 SE 13.7 fl oz at V12	10.0 i	23.7 h	21.0 de	57.5 bc	6.8 ef
Miravis Neo 2.5 SE 13.7 fl oz at R1	28.0 a-d	80.0 abc	23.8 abc	50.0 def	4.8 ef
Miravis Neo 2.5 SE 13.7 fl oz at V12 fb R3	3.8 j	36.7 gh	5.5 g	66.3 a	50.0 a
Trivapro 2.21 SE 13.7 fl oz at R1	28.3 abc	85.0 ab	24.5 ab	48.8 ef	3.3 f
Delaro Complete 458 SC 8.0 fl oz at R1	21.2 fg	60.0 b-f	24.8 a	55.0 bcd	5.5 ef
Veltyma 3.34 S 7.0 fl oz at R1	27.3 a-e	79.8 abc	23.8 abc	50.0 def	5.5 ef
Aproach Prima 2.34 SC 6.8 fl oz at R1	22.6 d-g	77.0 abc	23.3 abc	50.0 def	3.8 f
Brixen 15.0 fl oz at R1	22.3 efg	66.8 b-f	22.5 bcd	53.8 cde	5.8 ef
Brixen 13.0 fl oz at R1	21.3 fg	64.8 b-f	22.5 bcd	53.8 cde	7.0 ef
Brixen 10.0 fl oz at R1	17.8 gh	73.3 a-d	22.3 cd	52.5 cde	4.5 ef
Zolera ODX 5.0 fl oz at R1	24.0 a-f	69.3 a-d	19.9 e	53.8 cde	10.5 e
Vacciplant SL 14.0 fl oz at R1	29.3 ab	85.0 ab	25.0 a	45.0 fg	2.8 f
Zolera ODX 5.0 fl oz + Vacciplant SL 14.0 fl oz at R1	23.0 c-g	78.5 abc	20.0 e	52.5 cde	5.7 ef
Zolera ODX 5.0 fl oz at R2	14.1 hi	63.5 b-f	10.8 f	57.5 bc	27.5 d
Vacciplant SL 14 fl oz at R2	28.0 a-d	79.8 abc	23.3 abc	48.8 ef	4.0 f
Zolera ODX 5.0 fl oz + Vacciplant SL 14.0 fl oz at R2	10.9 i	47.8 efg	11.1 f	57.5 bc	33.8 c
Veltyma 3.34 S 7.0 fl oz at R2	14.6 hi	44.5 fgh	6.9 g	60.0 b	47.5 a
Delaro Complete 458 SC 8.0 fl oz at R2	13.7 hi	51.5 d-g	10.0 f	60.0 b	40.0 b
Nontreated control	31.8 a	85.0 ab	25.3 a	45.0 fg	2.0 f
P-value[v]	*0.0001*	*0.0001*	*0.0001*	*0.0001*	*0.0001*

[z] Fungicides were applied at V12, silk (R1), blister (R2), and milk (R3) growth stages on August 2, August 6, August 20, and August 30, respectively. All treatments applied at R1, R2, and R3 contained a nonionic surfactant (Preference) at a rate of 0.25% v/v. fb = followed by.

[y] Tar spot stroma was visually assessed as a percentage (0–100%) of leaf area on five plants in each plot at the ear leaf.

[x] Tar spot chlorotic and necrotic symptoms were visually assessed as a percentage (0–100%) of leaf area on five plants in each plot at the ear leaf.

[w] Canopy greenness was visually assessed as a percentage (0–100%) of crop canopy green on September 22 and September 29.

[v] A generalized linear mixed model analysis of variance was performed using PROC GLIMMIX. Values are least squares means, and values with different letters are significantly different based on least squares means test (α=0.05).

TABLE 37. *Effect of Fungicide on Lodging and Yield of Corn*

TREATMENT, RATE/ACRE, AND TIMING[z]	LODGING %[y]	HARVEST MOISTURE %	TEST WEIGHT LB/BU	YIELD[x] BU/ACRE
Nontreated control	0.5 a	20.5 hi	52.8	149.0 fg
Miravis Neo 2.5 SE 13.7 fl oz at V12	0.0 b	21.5 e-h	53.8	161.7 c-f
Miravis Neo 2.5 SE 13.7 fl oz at R1	0.0 b	20.8 ghi	53.7	155.1 d-g
Miravis Neo 2.5 SE 13.7 fl oz at V12 fb R3	0.0 b	23.9 a	52.5	181.5 ab
Trivapro 2.21 SE 13.7 fl oz at R1	0.5 a	21.2 f-i	55.4	160.5 c-g
Delaro Complete 458 SC 8.0 fl oz at R1	0.0 b	22.6 b-e	52.6	153.0 efg
Veltyma 3.34 S 7.0 fl oz at R1	0.0 b	22.3 b-f	52.3	161.9 c-f
Aproach Prima 2.34 SC 6.8 fl oz at R1	0.0 b	21.7 e-h	53.0	149.7 fg
Brixen 15.0 fl oz at R1	0.0 b	22.2 b-f	52.5	158.0 c-g
Brixen 13.0 fl oz at R1	0.0 b	21.7 d-h	53.1	151.5 efg
Brixen 10.0 fl oz at R1	0.0 b	21.0 f-i	53.6	156.5 c-g
Zolera ODX 5.0 fl oz at R1	0.0 b	21.8 d-g	53.1	164.4 cde
Vacciplant SL 14.0 fl oz at R1	0.0 b	21.2 f-i	52.4	151.9 efg
Zolera ODX 5.0 fl oz + Vacciplant SL 14.0 fl oz at R1	0.0 b	21.2 f-i	53.2	158.6 c-g
Zolera ODX 5.0 fl oz at R2	0.0 b	23.2 abc	53.2	170.8 bc
Vacciplant SL 14 fl oz at R2	0.0 b	22.1 c-f	52.7	151.8 efg
Zolera ODX 5.0 fl oz + Vacciplant SL 14.0 fl oz at R2	0.0 b	23.3 abc	52.5	169.1 bcd
Veltyma 3.34 S 7.0 fl oz at R2	0.0 b	23.0 a-d	53.4	188.7 a
Delaro Complete 458 SC 8.0 fl oz at R2	0.0 b	23.5 ab	52.9	181.8 ab
Nontreated control	0.5 a	20.5 hi	52.8	149.0 fg
P-value[w]	*0.0055*	*0.0001*	*0.1689*	*0.0001*

[z] Fungicides were applied at V12, silk (R1), blister (R2), and milk (R3) growth stages on August 2, August 6, August 20, and August 30, respectively. All treatments applied at R1, R2, and R3 contained a nonionic surfactant (Preference) at a rate of 0.25% v/v. fb = followed by.

[y] Lodging was assessed as a percentage of lodged stalks when pushed from shoulder height to 45° from vertical on September 29.

[x] Yields were adjusted to 15.5% moisture after harvesting on November 3.

[w] All data were analyzed in SAS 9.4 (SAS Institute, Cary, NC). A generalized linear mixed model analysis of variance was performed using PROC GLIMMIX. Values are least squares means, and values with different letters are significantly different based on least squares means test (α=0.05).

EVALUATION OF XYWAY AND FOLIAR FUNGICIDE PROGRAMS FOR TAR SPOT IN CORN IN NORTHWESTERN INDIANA, 2021 (COR21-16.PPAC)

S. Shim and D. E. P. Telenko, Department of Botany and Plant Pathology, Purdue University West Lafayette, IN 47907-2054

CORN (*ZEA MAYS* W2585SSRIB)

Tar spot, *Phyllachora maydis*

A trial was established at the Pinney Purdue Agricultural Center (PPAC) in Porter County, Indiana. The experiment was a randomized complete block design with four replications. Plots were 10 feet wide and 30 feet long and consisted of four rows, and the two center rows were used for evaluation. The previous crop was corn. Standard practices for grain corn production in Indiana were followed. Corn hybrid W2585SSRIB was planted in 30-inch row spacing at a rate of 2 seeds/foot on May 27. In-furrow treatments was applied at planting at 10 gal/acre. All foliar fungicide applications were applied at 15 gal/acre and 40 psi using a Lee self-propelled sprayer equipped with a 10-foot boom, fitted with six TJ-VS 8002 nozzles spaced 20 inches apart. Fungicides were applied on July 23, August 6, and August 30 at V10, silk (R1) and milk (R3) growth stages, respectively. Disease ratings were assessed on September 28 at maturity (R6) growth stage. Tar spot was rated by visually assessing the percentage of stroma per leaf on five plants in each plot at the ear leaf. Values for each plot were averaged before analysis. The two center rows of each plot were harvested on November 4, and yields were adjusted to 15.5% moisture. All data were analyzed in SAS 9.4 (SAS Institute, Cary, NC). A generalized linear mixed model analysis of variance was performed using PROC GLIMMIX. Values are least squares means, and values with different letters are significantly different based on least squares means test (α=0.05).

In 2021, weather conditions were favorable for disease. Tar spot was the most prominent disease in the trial and reached high severity. There was no significant effect on treatment for tar spot stroma severity on September 28 (Table 38). On September 28 at R6, treatments that included Topguard at V10 or R3 were the only plots greener than the nontreated control. There was no significant effect of treatment on harvest moisture, test weight, and yield of corn.

TABLE 38. *Effect of Fungicide on Tar Spot, Canopy Greenness, and Yield of Corn*

TREATMENT, RATE/ACRE, AND TIMING[z]	TAR SPOT[y] %	CANOPY[x] %	HARVEST MOISTURE %	TEST WEIGHT LB/BU	YIELD[w] BU/ACRE
Nontreated control	25.8	17.5 c	52.2	31.8	167.5
Xyway LFR 15.2 fl oz in-furrow	25.3	21.3 c	51.8	33.2	172.9
Xyway LFR 10.5 fl oz in-furrow fb Topguard EQ 4.29 5.0 fl oz at V10	20.0	31.3 b	52.9	33.4	172.4
Xyway LFR 10.5 fl oz in-furrow fb Topguard EQ 4.29 5.0 fl oz at R1	24.8	17.5 c	51.7	31.6	165.5
Xyway LFR 10.5 fl oz in-furrow fb Topguard EQ 4.29 5.0 fl oz at R3	17.2	42.5 a	52.4	32.5	172.3
Topguard EQ 4.29 5.0 fl oz at R1	23.3	18.8 c	52.3	31.8	167.9
Xyway LFR 15.2 fl oz 2x2 at plant	26.0	18.8 c	52.4	33.8	173.8
Xyway LFR 10.5 fl oz in-furrow	25.0	21.3 c	52.5	32.3	157.6
Trivapro 2.21 SE 13.7 fl oz at R1	23.0	22.5 c	53.2	32.5	171.6
Xyway LFR 15.2 fl oz in-furrow fb Trivapro 2.21 SE 13.7 fl oz at R1	22.8	20.0 c	52.3	35.0	130.7
P-value[v]	0.1087	0.0001	0.3831	0.3709	0.5565

[z] In-furrow treatments were applied at planting at 10 gal/acre. Fungicides were applied on July 23, August 6, and August 20 at V10, silk (R1), and milk (R3) growth stages, respectively. fb = followed by.

[y] Tar spot stroma was visually assessed as a percentage (0–100%) of leaf area on five plants in each plot at the ear leaf on September 28.

[x] Canopy greenness was visually assessed as a percentage (0–100%) of crop canopy green on September 28.

[w] Yields were adjusted to 15.5% moisture after harvesting on November 4.

[v] All data were analyzed in SAS 9.4 (SAS Institute, Cary, NC). A generalized linear mixed model analysis of variance was performed using PROC GLIMMIX. Values are least squares means, and values with different letters are significantly different based on least squares means test (α=0.05).

EVALUATION OF XYWAY PROGRAMS IN CORN FOR TAR SPOT IN NORTHWESTERN INDIANA, 2021 (COR21-21.PPAC)

S. Shim and D. E. P. Telenko, Department of Botany and Plant Pathology, Purdue University West Lafayette, IN 47907-2054

CORN (*ZEA MAYS* W2585SSRIB)

Tar spot, *Phyllachora maydis*

A trial was established at the Pinney Purdue Agricultural Center (PPAC) in Porter County, Indiana. The experiment was a randomized complete block design with six replications. Plots were 10 feet wide and 30 feet long and consisted of four rows, and the two center rows were used for evaluation. The previous crop was corn. Standard practices for grain corn production in Indiana were followed. Corn hybrid W2585SSRIB was planted in 30-inch row spacing at a rate of 2 seeds/foot on May 27. Standard practices for nonirrigated grain corn production in Indiana were followed. In-furrow treatments was applied at planting at 10 gal/acre. All foliar fungicide applications were applied at 15 gal/acre and 40 psi using a Lee self-propelled sprayer equipped with a 10-foot boom, fitted with six TJ-VS 8002 nozzles spaced 20 inches apart. Foliar fungicides were applied on August 8 at the silk (R1) growth stage. Disease ratings were assessed on September 14 and September 28 at dent (R5) growth stages. Tar spot was rated by visually assessing the percentage of stroma per leaf on five plants in each plot at the ear leaf. Values for each plot were averaged before analysis. The two center rows of each plot were harvested on November 4, and yields were adjusted to 15.5% moisture. All data were analyzed in SAS 9.4 (SAS Institute, Cary, NC). A generalized linear mixed model analysis of variance was performed using PROC GLIMMIX. Values are least squares means, and values with different letters are significantly different based on least squares means test (α=0.05).

In 2021, weather conditions were favorable for disease. Tar spot was the most prominent disease in the trial and reached high severity. Xyway in-furrow followed by Topguard at R1 and Veltyma at R1 significantly reduced the percent of tar spot stroma severity over the nontreated control on September 14 and September 28 (Table 39). There were no significant differences between treatments and the nontreated control for percent of canopy green, harvest moisture, test weight, and corn yield.

TABLE 39. *Effect of Fungicide on Tar Spot, Canopy Greenness, and Yield of Corn*

TREATMENT, RATE/ACRE, AND TIMING[z]	TAR SPOT[y] % SEP 14	TAR SPOT[y] % SEP 28	CANOPY[x] % SEP 28	HARVEST MOISTURE %	TEST WEIGHT LB/BU	YIELD[w] BU/ACRE
Nontreated control	9.8 a	0.3	21.5	51.6	163.8	18.3
Xyway LFR 15.2 fl oz in-furrow	7.5 b	0.0	21.3	51.9	168.0	17.5
Xyway LFR 10.5 fl oz in-furrow fb Topguard EQ 4.29 5.0 fl oz at R1	5.8 bc	0.0	21.5	52.6	168.9	21.7
Topguard EQ 4.29 5.0 fl oz at R1	7.3 bc	0.0	21.7	52.0	166.0	18.3
Veltyma 3.34 S 7.0 fl oz at R1	5.4 c	0.0	22.2	51.4	160.1	20.0
P-value[v]	0.0022	0.0751	0.6212	0.4205	0.5578	0.3843

[z] Xyway was applied in-furrow at planting on May 27. Topguard and Veltyma were applied on August 8 at the silk (R1) growth stage and contained a nonionic surfactant (Preference) at a rate of 0.25% v/v. fb = followed by.

[y] Tar spot stroma was visually assessed as a percentage (0–100%) of leaf area on five plants in each plot at the ear leaf on September 14 and September 28.

[x] Canopy greenness was visually assessed as a percentage (0–100%) of canopy green on September 28.

[w] Yields were adjusted to 15.5% moisture after harvesting on November 4.

[v] All data were analyzed in SAS 9.4 (SAS Institute, Cary, NC). A generalized linear mixed model analysis of variance was performed using PROC GLIMMIX. Values are least squares means, and values with different letters are significantly different based on least squares means test (α=0.05).

FUNGICIDE COMPARISON FOR TAR SPOT IN CORN IN NORTHWESTERN INDIANA, 2021 (COR21-23.PPAC)

S. Shim and D. E. P. Telenko, Department of Botany and Plant Pathology, Purdue University West Lafayette, IN 47907-2054

CORN (*ZEA MAYS* W2585SSRIB)

Tar spot, *Phyllachora maydis*

A trial was established at the Pinney Purdue Agricultural Center (PPAC) in Porter County, Indiana. The experiment was a randomized complete block design with four replications. Plots were 10 feet wide and 30 feet long and consisted of four rows, and the two center rows were used for evaluation. The previous crop was corn. Standard practices for grain corn production in Indiana were followed. Corn hybrid W2585SSRIB was planted in 30-inch row spacing at a rate of 34,000 seeds/acre on May 25. The field was overhead irrigated weekly at 1 inch unless weekly rainfall was 1 inch or higher to encourage disease. In-furrow fungicides was applied at planting at 10 gal/acre. All foliar fungicide applications were applied at 15 gal/acre and 40 psi using either a CO_2 backpack sprayer or a Lee self-propelled sprayer equipped with a 10-foot boom, fitted with six TJ-VS 8002 nozzles spaced 20 inches apart. Fungicides were applied on May 25 in-furrow and on July 9, August 2, August 6, and August 30 at V5, V12, silk (R1), and milk (R3) growth stages, respectively. Disease ratings were assessed on September 22 and September 29. Tar spot was rated by visually assessing the percentage of stroma per leaf on five plants in each plot at the ear leaf. Values for each plot were averaged before analysis. The two center rows of each plot were harvested on November 3, and yields were adjusted to 15.5% moisture. All data were analyzed in SAS 9.4 (SAS Institute, Cary, NC). A generalized linear mixed model analysis of variance was performed using PROC GLIMMIX. Values are least squares means, and values with different letters are significantly different based on least squares means test (α=0.05).

In 2021, weather conditions were favorable for disease. Tar spot was the most prominent disease in the trial and reached high severity. There were significant differences between fungicide treatments and the nontreated control for all disease ratings. On September 22 and September 29 at R6, Veltyma at V12 and Veltyma applied at V12 followed by (fb) R3 were the only treatments with reduced tar spot stromata (Table 40). Vetlyma applied at V12 fb R3 increased percent canopy green as compared to the nontreated control. The programs that included Veltyma at V12 followed by R3 has significantly higher grain moisture and yield than the nontreated control and other treatments except for grain moisture with Priaxor fb Veltyma. There was no significant effect of treatment on test weight.

TABLE 40. *Effect of Fungicide Treatment on Tar Spot, Canopy Greenness, and Yield of Corn*

TREATMENT, RATE/ACRE, AND TIMING[z]	TAR SPOT[y] % SEP 22	TAR SPOT[y] % SEP 29	CANOPY[x] % SEP 29	HARVEST MOISTURE %	TEST WEIGHT LB/BU	YIELD[w] BU/ACRE
Nontreated control	30.8 a	24.8 a	7.8 b	20.5 b	52.9	168.1 b
Headline 2.08 SC 6.0 fl oz in-furrow	31.3 a	25.0 a	6.5 b	20.0 b	54.2	167.0 b
Priaxor 4.17 SC 4.0 fl oz at V5	31.5 a	24.8 a	4.3 b	20.3 b	53.9	172.5 b
Veltyma 3.34 S 7.0 fl oz at V12	14.4 b	17.8 b	14.5 b	20.5 b	54.4	182.4 b
Veltyma 3.34 S 7.0 fl oz at R1	30.5 a	24.8 a	18.8 ab	20.1 b	54.1	172.7 b
Veltyma 3.3.4 S 7.0 fl oz at V12 fb Veltyma 3.3.4 S 7.0 fl oz at R3	4.0 c	4.1 c	39.5 a	23.2 a	53.6	199.4 a
Priaxor 4.17 SC 4.0 fl oz at V5 fb Veltyma 3.3.4 S 7.0 fl oz at R1	25.5 a	22.1 a	6.8 b	21.7 ab	53.1	171.3 b
P-value[v]	0.0001	0.0001	0.0768	0.0408	0.5059	0.0066

[z] In-furrow treatments were applied at planting on May 25. Foliar fungicide treatments were applied on May 25 in-furrow and on July 9, August 2, August 6, and August 30 at V5, V12, silk (R1), milk (R3) growth stages, respectively. Foliar fungicide treatments at R1 and R3 contained a nonionic surfactant (Preference) at a rate of 0.25% v/v. fb = followed by.

[y] Tar spot stroma was visually assessed as a percentage (0–100%) of leaf area on five plants in each plot at the ear leaf on September 22 and September 29.

[x] Canopy greenness was visually assessed as a percentage (0–100%) of crop canopy green on September 29.

[w] Yields were adjusted to 15.5% moisture after harvesting on November 3.

[v] All data were analyzed in SAS 9.4 (SAS Institute, Cary, NC). A generalized linear mixed model analysis of variance was performed using PROC GLIMMIX. Values are least squares means, and values with different letters are significantly different based on least squares means test (α=0.05).

FUNGICIDE COMPARISON FOR TAR SPOT IN CORN IN NORTHWESTERN INDIANA, 2021 (COR21-27.PPAC)

S. Shim and D. E. P. Telenko, Department of Botany and Plant Pathology, Purdue University West Lafayette, IN 47907-2054

CORN (*ZEA MAYS* W2585SSRIB)

Tar spot, *Phyllachora maydis*

A trial was established at the Pinney Purdue Agricultural Center (PPAC) in Porter County, Indiana. The experiment was a randomized complete block design with four replications. Plots were 10 feet wide and 30 feet long and consisted of four rows, and the two center rows were used for evaluation. The previous crop was corn. Standard practices for grain corn production in Indiana were followed. Corn hybrid W2585SS-RIB was planted in 30-inch row spacing at a rate of 34,000 seeds/acre on May 25. The field was overhead irrigated weekly at 1 inch unless weekly rainfall was 1 inch or higher to encourage disease. In-furrow fungicides were applied at planting in 10 gal/acre. All foliar fungicide applications were applied at 15 gal/acre and 40 psi using a Lee self-propelled sprayer equipped with a 10-foot boom, fitted with six TJ-VS 8002 nozzles spaced 20 inches apart. Fungicides were applied on August 6 at beginning bloom (R1) growth stage. Disease ratings were assessed on September 22 and September 29 at dent (R5) and maturity (R6) growth stages, respectively. Tar spot was rated by visually assessing the percentage of stroma per leaf on five plants in each plot at the ear leaf. Values for each plot were averaged before analysis. The two center rows of each plot were harvested on November 3, and yields were adjusted to 15.5% moisture. All data were analyzed in SAS 9.4 (SAS Institute, Cary, NC). A generalized linear mixed model analysis of variance was performed using PROC GLIMMIX. Values are least squares means, and values with different letters are significantly different based on least squares means test ($\alpha=0.05$).

In 2021, weather conditions were favorable for disease. Tar spot was the most prominent disease in the trial and reached high severity. On September 22 at R5 growth stage, all fungicide treatments reduced tar spot stroma severity compared to the nontreated control except Xyway and Lucento (Table 41). Delaro Complete significantly reduced tar spot stroma severity compared to the nontreated control and other fungicide treatments on September 22. There was no significant difference between treatments for tar spot stroma severity on September 29 at R6 growth stage. There was no significant effect of treatment on percent canopy greenness, moisture, test weight, and yield of corn.

TABLE 41. *Effect of Fungicide Treatment on Tar Spot, Canopy Greenness, and Yield of Corn*

TREATMENT, RATE/ACRE, AND TIMING[z]	TAR SPOT[y] % SEP 22	TAR SPOT[y] % SEP 29	CANOPY GREEN[x] %	HARVEST MOISTURE %	TEST WEIGHT LB/BU	YIELD[w] BU/ACRE
Nontreated control	24.3 a	25.5	7.0	19.6	55.0	177.7
Veltyma 3.34 S 7.0 fl oz at R1	17.5 b	24.5	4.0	19.8	54.3	177.6
Xyway LFR 15.2 fl oz in-furrow	23.0 a	25.3	4.5	19.4	54.5	171.6
Delaro Complete 458 SC 8.0 fl oz at R1	12.3 c	23.3	10.0	20.1	54.9	183.0
Lucento 4.1 SC 5.0 fl oz at R1	20.5 ab	24.8	8.5	19.8	54.1	181.5
Miravis Neo 2.5 SE 13.6 fl oz at R1	18.9 b	24.0	12.5	19.8	54.6	180.4
Trivapro 2.21 SE 13.7 fl oz at R1	18.0 b	24.8	6.3	20.1	54.5	180.3
P-value[v]	0.0001	0.2952	0.1042	0.7936	0.3406	0.9237

[z] In-furrow treatments were applied at planting on May 25. Foliar fungicide treatments were applied on August 6 at silk (R1) growth stage. All foliar fungicide treatments contained a nonionic surfactant (Preference) at a rate of 0.25% v/v.

[y] Tar spot stroma was visually assessed as a percentage (0–100%) of leaf area on five plants in each plot at the ear leaf on September 29.

[x] Canopy greenness was visually assessed as a percentage (0–100%) of crop canopy green on September 29.

[w] Yields were adjusted to 15.5% moisture after harvesting on November 3.

[v] All data were analyzed in SAS 9.4 (SAS Institute, Cary, NC). A generalized linear mixed model analysis of variance was performed using PROC GLIMMIX. Values are least squares means, and values with different letters are significantly different based on least squares means test (α=0.05).

FUNGICIDE COMPARISON FOR TAR SPOT IN CORN IN NORTHWESTERN INDIANA, 2021 (COR21-29.PPAC)

S. Shim and D. E. P. Telenko, Department of Botany and Plant Pathology, Purdue University West Lafayette, IN 47907-2054

CORN (*ZEA MAYS* W2585SSRIB)

Tar spot, *Phyllachora maydis*

A trial was established at the Pinney Purdue Agricultural Center (PPAC) in Porter County, Indiana. The experiment was a randomized complete block design with four replications. Plots were 10 feet wide and 30 feet long and consisted of four rows, and the two center rows were used for evaluation. The previous crop was corn. Standard practices for grain corn production in Indiana were followed. Corn hybrid W2585SSRIB was planted in 30-inch row spacing at a rate of 34,000 seeds/acre on May 27. All foliar fungicide applications were applied at 15 gal/acre and 40 psi using a Lee self-propelled sprayer equipped with a 10-foot boom, fitted with six TJ-VS 8002 nozzles spaced 20 inches apart. Fungicides were applied on August 6 at silk (R1) growth stage. Disease ratings were assessed on September 28 at maturity (R6) growth stage. Tar spot was rated by visually assessing the percentage of stroma per leaf on five plants in each plot at the ear leaf. Values for each plot were averaged before analysis. The two center rows of each plot were harvested on November 3, and yields were adjusted to 15.5% moisture. All data were analyzed in SAS 9.4 (SAS Institute, Cary, NC). A generalized linear mixed model analysis of variance was performed using PROC GLIMMIX. Values are least squares means, and values with different letters are significantly different based on least squares means test (α=0.05).

In 2021, weather conditions were favorable for disease. Tar spot was the most prominent disease in the trial and reached high severity. There were no significant differences between fungicide treatments and the nontreated control for disease ratings (Tables 42). All fungicide treatments increased percent of canopy greenness over the nontreated control on September 28. There was no significant effect of treatment on moisture, test weight, and yield of corn.

TABLE 42. *Effect of Fungicide on Canopy Greenness and Yield of Corn*

TREATMENT AND RATE/ACRE[z]	TAR SPOT[y] % SEP 14	TAR SPOT[y] % SEP 28	CANOPY[x] % SEP 28	HARVEST MOISTURE %	TEST WEIGHT LB/BU	YIELD[w] BU/ACRE
Nontreated control	8.4	23.5	18.8 b	22.1	52.2	143.3
Delaro Complete 458 SC 8.0 fl oz	3.9	20.3	33.8 a	22.0	52.6	177.5
Delaro Complete 458 SC 12.0 fl oz	4.0	22.5	32.5 a	21.8	53.0	168.5
Delaro 325 SE 11.8 fl oz	5.4	21.8	32.5 a	22.2	52.6	171.7
Veltyma 3.34S 7.0 fl oz	4.2	22.0	37.5 a	22.5	52.7	151.2
Miravis Neo 2.5 SE 13.7 fl oz	4.6	21.0	30.0 a	21.7	52.7	166.8
Trivapro 2.21 SE 13.7 fl oz	5.2	22.0	35.0 a	23.1	51.9	153.8
P-value[v]	0.1277	0.3662	0.0124	0.2167	0.4903	0.4870

[z] Foliar fungicide treatments were applied on August 6 at silk (R1) growth stages. Fungicide treatments contained a nonionic surfactant (Preference) at a rate of 0.25% v/v.

[y] Tar spot stroma was visually assessed as a percentage (0–100%) of leaf area on five plants in each plot at the ear leaf on September 14 and September 28.

[x] Canopy greenness was visually assessed as a percentage (0–100%) of crop canopy green on September 28.

[w] Yields were adjusted to 15.5% moisture after harvesting on November 3.

[v] All data were analyzed in SAS 9.4 (SAS Institute, Cary, NC). A generalized linear mixed model analysis of variance was performed using PROC GLIMMIX. Values are least squares means, and values with different letters are significantly different based on least squares means test (α=0.05).

FUNGICIDE COMPARISON FOR TAR SPOT IN CORN IN NORTHWESTERN INDIANA, 2021 (COR21-30.PPAC)

S. Shim and D. E. P. Telenko, Department of Botany and Plant Pathology, Purdue University West Lafayette, IN 47907-2054

CORN (*ZEA MAYS* W2585SSRIB)

Tar spot, *Phyllachora maydis*

A trial was established at the Pinney Purdue Agricultural Center (PPAC) in Porter County, Indiana. The experiment was a randomized complete block design with four replications. Plots were 10 feet wide and 30 feet long and consisted of four rows, and the two center rows were used for evaluation. The previous crop was corn. Standard practices for grain corn production in Indiana were followed. Corn hybrid W2585SSRIB was planted in 30-inch row spacing at a rate of 34,000 seeds/acre on May 27. The field was overhead irrigated weekly at 1 inch unless weekly rainfall was 1 inch or higher to encourage disease. All foliar fungicide applications were applied at 15 gal/acre and 40 psi using a Lee self-propelled sprayer equipped with a 10-foot boom, fitted with six TJ-VS 8002 nozzles spaced 20 inches apart. Fungicides were applied on August 2 and August 6 at V14 and silk (R1) growth stages, respectively. Disease ratings were assessed on September 22 and September 29 at dent (R5) and maturity (R6) growth stages, respectively. Tar spot was rated by visually assessing the percentage of stroma per leaf on five plants in each plot at the ear leaf. Values for each plot were averaged before analysis. The two center rows of each plot were harvested on November 3, and yields were adjusted to 15.5% moisture. All data were analyzed in SAS 9.4 (SAS Institute, Cary, NC). A generalized linear mixed model analysis of variance was performed using PROC GLIMMIX. Values are least squares means, and values with different letters are significantly different based on least squares means test (α=0.05).

In 2021, weather conditions were favorable for disease. Tar spot was the most prominent disease in the trial and reached high severity. There were no significant differences between fungicide treatments and the nontreated control for tar spot stroma severity on September 22 and September 29 (Table 43). There was no significant effect of treatment on percent canopy greenness, moisture, test weight, and yield of corn.

TABLE 43. *Effect of Fungicide on Canopy Greenness and Yield of Corn*

TREATMENT, RATE/ACRE, AND TIMING[z]	TAR SPOT[y] % SEP 22	TAR SPOT[y] % SEP 29	CANOPY[x] % SEP 29	HARVEST MOISTURE %	TEST WEIGHT LB/BU	YIELD[w] BU/ACRE
Nontreated control	30.3	25.3	2.5	19.9	53.9	158.6
Headline AMP 1.68 SE 10.0 fl oz at V14	26.5	24.8	5.0	20.9	53.0	160.8
Quilt XCEL 2.2 SE 10.5 fl oz at V14	30.5	24.8	4.3	20.9	53.6	141.9
Veltyma 3.24 S 7.0 fl oz at V14	16.6	21.8	10.5	21.3	53.0	167.4
Miravis Neo 2.5 SE 13.7 fl oz at V14	22.9	24.5	5.5	21.5	54.2	170.5
Headline AMP 1.68 SE 6.0 fl oz at R1	30.3	24.3	4.5	19.7	53.6	152.4
Quadris 2.1 F 9.0 fl oz at R1	29.4	23.3	4.0	20.6	53.6	156.1
Headline AMP 1.68 SE 10.0 fl oz at R1	24.3	23.8	3.5	20.8	71.3	152.9
Quilt XCEL 2.2 SE 10.5 fl oz at R1	30.8	24.8	3.8	20.4	53.7	157.8
Veltyma 3.24 S 7.0 fl oz at R1	23.5	24.3	19.8	21.0	53.2	154.9
Miravis Neo 2.5 SE 13.7 fl oz at R1	27.5	24.3	3.8	20.1	54.1	150.0
P-value[v]	*0.0919*	*0.5970*	*0.4663*	*0.4283*	*0.4197*	*0.4119*

[z] Foliar fungicide treatments were applied on August 2 and August 6 at V14 and silk (R1) growth stages, respectively. Foliar fungicide treatments at R1 contained a nonionic surfactant (Preference) at a rate of 0.25% v/v.

[y] Tar spot stroma was visually assessed as a percentage (0–100%) of leaf area on five plants in each plot at the ear leaf on September 22 and September 29 at dent (R5) and maturity (R6) growth stages, respectively.

[x] Canopy greenness was visually assessed as a percentage (0–100%) of crop canopy green on September 29.

[w] Yields were adjusted to 15.5% moisture after harvesting on November 3.

[v] All data were analyzed in SAS 9.4 (SAS Institute, Cary, NC). A generalized linear mixed model analysis of variance was performed using PROC GLIMMIX. Values are least squares means, and values with different letters are significantly different based on least squares means test ($\alpha=0.05$).

EVALUATION OF VELTYMA TIMING PROGRAMS FOR TAR SPOT IN CORN IN NORTHWESTERN INDIANA, 2021 (COR21-35.PPAC)

S. Shim and D. E. P. Telenko, Department of Botany and Plant Pathology, Purdue University West Lafayette, IN 47907-2054

CORN (*ZEA MAYS* W2585SSRIB)

Tar spot, *Phyllachora maydis*

A trial was established at the Pinney Purdue Agricultural Center (PPAC) in Porter County, Indiana. The experiment was a randomized complete block design with four replications. Plots were 10 feet wide and 30 feet long and consisted of four rows, and the two center rows were used for evaluation. The previous crop was corn. Standard practices for grain corn production in Indiana were followed. Corn hybrid W2585SSRIB was planted in 30-inch row spacing at a rate of 2 seeds/foot on May 27. In-furrow treatments was applied at planting at 10 gal/acre. All foliar fungicide applications were applied at 15 gal/acre and 40 psi using a Lee self-propelled sprayer equipped with a 10-foot boom, fitted with six TJ-VS 8002 nozzles spaced 20 inches apart. Fungicides were applied on July 23, August 2, August 6, August 20, August 30, September 10, and September 16 at V8, V12, silk (R1), blister (R2), milk (R3), dough (R4), and dent (R5) growth stages, respectively. Disease ratings were assessed on September 28 at maturity (R6) growth stage. Tar spot was visually assessed as a percentage of stroma per leaf on five plants in each plot at the ear leaf. Values for each plot were averaged before analysis. The two center rows of each plot were harvested on November 4, and yields were adjusted to 15.5% moisture. All data were analyzed in SAS 9.4 (SAS Institute, Cary, NC). A generalized linear mixed model analysis of variance was performed using PROC GLIMMIX. Values are least squares means, and values with different letters are significantly different based on least squares means test (α=0.05).

In 2021, weather conditions were favorable for disease. Tar spot was the most prominent disease in the trial and reached high severity. Tar spot stroma severity on all leaves on September 28 was significantly reduced over the nontreated by all fungicide programs except Veltyma applied at R4 and R5 (Table 44). Veltyma applied at R2 resulted in the lowest amount of tar spot stroma on September 28. Canopy greenness was increased by treatments applied at R2, R3, and R1 followed by R4. Veltyma applied at R3 significantly increased yield over the nontreated controls.

TABLE 44. *Effect of Fungicide on Tar Spot Stroma Severity, Canopy Greenness, and Yield of Corn*

TREATMENT, RATE/ACRE, AND TIMING[z]	TAR SPOT[y] %	CANOPY[x] %	HARVEST MOISTURE %	TEST WEIGHT LB/BU	YIELD[w] BU/ACRE
Nontreated control	25.5 a	26.3 e	20.6 c	52.9	163.4 c
Nontreated control	24.0 ab	32.5 de	20.7 c	54.0	178.7 bc
Veltyma 3.34 S 7.0 fl oz at V8	15.5 e	41.3 bcd	22.2 abc	53.0	180.9 bc
Veltyma 3.34 S 7.0 fl oz at V12	16.0 de	31.3 de	21.5 bc	53.4	163.5 c
Veltyma 3.34 S 7.0 fl oz at R1	19.5 cd	40.0 cd	21.2 bc	53.4	167.1 c
Veltyma 3.34 S 7.0 fl oz at R2	10.2 f	53.8 ab	23.3 a	53.5	205.4 ab
Veltyma 3.34 S 7.0 fl oz at R3	14.6 e	55.0 a	22.5 ab	53.4	209.3 a
Veltyma 3.34 S 7.0 fl oz at R4	21.5 bc	42.5 a-d	22.8 ab	52.6	159.1 c
Veltyma 3.34 S 7.0 fl oz at R5	22.5 abc	35.0 cde	21.4 bc	52.9	177.0 c
Veltyma 3.34 S 7.0 fl oz at R1 fb R4	16.0 de	46.3 abc	22.2 abc	53.4	173.7 c
P-value[v]	*0.0001*	*0.0018*	*0.0446*	*0.5876*	*0.0036*

[z] Fungicides were applied on July 23, August 2, August 6, August 20, August 30, September 10, and September 16 at V8, V12, silk (R1), blister (R2), milk (R3), dough (R4), and dent (R5) growth stages, respectively. fb = followed by.

[y] Tar spot stroma was visually assessed as a percentage (0–100%) of leaf area on five plants in each plot at the ear leaf on September 28.

[x] Canopy greenness was visually assessed as a percentage (0–100%) of crop canopy green on September 28.

[w] Yields were adjusted to 15.5% moisture after harvesting on November 4.

[v] All data were analyzed in SAS 9.4 (SAS Institute, Cary, NC). A generalized linear mixed model analysis of variance was performed using PROC GLIMMIX. Values are least squares means, and values with different letters are significantly different based on least squares means test (α=0.05).

FUNGICIDE EVALUATION FOR WHITE MOLD IN SOYBEAN IN NORTHWESTERN INDIANA, 2021 (SOY21-02.PPAC)

A. Toogood, S. Shim, and D. E. P. Telenko, Department of Botany and Plant Pathology, Purdue University West Lafayette, IN 47907-2054

SOYBEAN (*GLYCINE MAX* P35T15E)

White mold, *Sclerotinia sclerotiorum*

A trial was established at the Pinney Purdue Agricultural Center (PPAC) in Porter County, Indiana. The experiment was a randomized complete block design with four replications. Plots were 6.7 feet wide and 30 feet long and consisted of four rows, and the two center rows were used for evaluation. The previous crop was sunflower. Standard practices for soybean production in Indiana were followed. Soybean cultivar P35T15E was planted in 20-inch row spacing at a rate of 8 seeds/foot on May 24. Inoculum of *S. sclerotiorum* was applied on the seedbed at 1.25 g/foot at planting. The field was overhead irrigated weekly at 1 inch unless weekly rainfall was 1 inch or higher to encourage disease. All pesticide applications were applied at 15 gal/acre and 40 psi using a CO_2 backpack sprayer equipped with a 10-foot boom, fitted with six TJ-VS 8002 nozzles spaced 20 inches apart. Fungicide treatments were applied on July 24, July 21, and July 30 at beginning bloom (R1), full bloom (R2), and beginning pod (R3) growth stages, respectively. Disease ratings were assessed on September 8 at full seed (R6) growth stage. White mold disease was assessed by counting the number of plants in each plot with symptoms. The two center rows of each plot were harvested on October 1, and yields were adjusted to 13% moisture. All data were analyzed in SAS 9.4 (SAS Institute, Cary, NC). A generalized linear mixed model analysis of variance was performed using PROC GLIMMIX. Values are least squares means, and values with different letters are significantly different based on least squares means test (α=0.05).

In 2021 weather conditions were not favorable for disease, and very little disease developed in the trial. White mold was present in the trial but only remained at low levels. There were differences between fungicide treatments and the nontreated control for disease ratings on September 8. White mold was not detected in the nontreated control plots but was found in Endura at R3, Delaro at R2, and Exp 1 at R2 plots at a low incidence (Table 45). There was no significant effect of treatment on moisture, test weight, and yield of soybean.

TABLE 45. *Effect of Fungicide on White Mold Incidence and Yield of Soybean*

TREATMENT, RATE/A, AND TIMING[z]	WHITE MOLD[y] #/PLOT	HARVEST MOISTURE %	TEST WEIGHT LB/BU	YIELD[x] BU/ACRE
Nontreated control	0.0 d	10.6	57.2	59.7
Endura 70 WDG 8.0 fl oz at R1 fb Endura 70 WDG 8.0 fl oz at R3	0.0 d	10.4	56.9	57.1
Endura 70 WDG 8.0 fl oz at R3	1.0 abc	10.6	57.2	59.3
Omega 16.0 fl oz at R3	0.5 bcd	10.4	57.5	54.5
Cobra 6.0 fl oz at R1	0.3 cd	10.7	57.4	56.2
Cobra 6.0 fl oz at R1 fb Domark 5.0 fl oz at R3	0.8 a-d	10.7	58.0	56.2
Omega 12.0 fl oz at R1 fb Miravis Neo 13.7 fl oz at R3	0.0 d	10.4	57.1	56.6
Delaro Complete 458 SC 8.0 fl oz at R2	1.5 a	11.0	60.5	58.8
Propulse 6.0 fl oz at R1 fb Delaro Complete 8.0 fl oz at R3	0.5 bcd	10.6	57.7	54.0
Miravis Neo 2.5 SE 16.0 fl oz at R2	0.3 cd	10.8	57.2	58.9
Exp A 13.7 fl oz at R2	1.3 ab	10.5	57.5	55.6
P-value[w]	0.0375	0.8200	0.1118	0.0790

[z] Fungicide treatments were applied on July 24, July 21, and July 30 at beginning bloom (R1), full bloom (R2), and beginning pod (R3) growth stages, respectively. All fungicide treatments contained a nonionic surfactant (Preference) at a rate of 0.25% v/v, no NIS with Cobra. All plots were inoculated with *S. sclerotiorum*. fb = followed by.

[y] White mold disease was assessed by counting the number of plants in each plot with symptoms on September 8.

[x] Yields were adjusted to 13% moisture after harvesting on October 1.

[w] All data were analyzed in SAS 9.4 (SAS Institute, Cary, NC). A generalized linear mixed model analysis of variance was performed using PROC GLIMMIX. Values are least squares means, and values with different letters are significantly different based on least squares means test (α=0.05).

FUNGICIDE COMPARISON FOR WHITE MOLD IN ORGANIC SOYBEAN IN NORTHWESTERN INDIANA, 2021 (SOY21-09.PPAC)

C. R. Da Silva, S. Shim, and D. E. P. Telenko, Department of Botany and Plant Pathology, Purdue University West Lafayette, IN 47907-2054

SOYBEAN (*GLYCINE MAX* DANE AND MN1410)

Frogeye leaf spot, *Cercospora sojina*
White mold, *Sclerotinia sclerotiorum*

A trial was established at the Pinney Purdue Agricultural Center (PPAC) in Porter County, Indiana. The experiment was a randomized complete block design with four replications. Plots were 6.7 feet wide and 30 feet long and consisted of four rows, and the two center rows were used for evaluation. The previous crop was sunflower. Cereal rye was planted on September 18, 2020, at a rate of 150 lbs/acre. On May 24 and May 25, the cover crop was terminated using either tillage or roller-crimping. Standard practices for soybean organic production in Indiana were followed. Organic soybean varieties Dane and MN1410 were planted in 20-inch row spacing at a rate of 8 seeds/foot on May 25. Inoculum of *S. sclerotiorum* was applied within the seedbed at 1.25 g/foot at planting, and 60 sclerotia per plot were spread between the middle two rows after tillage and before roller-crimping. The field was overhead irrigated weekly at 1 inch unless weekly rainfall was 1 inch or higher to encourage disease. All fungicides applications were applied at 15 gal/acre and 40 psi using a CO_2 backpack sprayer equipped with a 10-foot boom, fitted with four or six TJ-VS 8002 nozzles spaced 20 or 30 inches apart. Fungicides were applied on July 19 at full bloom (R2) growth stage. Disease ratings were assessed on August 26 at full seed (R6). Frogeye leaf spot (FLS) severity was rated by visually assessing the percentage (0–100%) of symptomatic leaf area in the upper canopy. The two center rows of each plot were harvested on September 28, and yields were adjusted to 13% moisture. All data were analyzed in SAS 9.4 (SAS Institute, Cary, NC). A generalized linear mixed model analysis of variance was performed using PROC GLIMMIX. Values are least squares means, and values with different letters are significantly different based on least squares means test (α=0.05).

In 2021, weather conditions were not favorable for disease. White mold was not observed in the plots. Frogeye leaf spot was the most prominent disease in the trial but only reached low severity. Main effects of cultivar, cover crop termination, and fungicide treatments are presented since there were no significant interactions between tillage, cultivar, and fungicide except for tillage by cultivar in yield (Table 46). Frogeye leaf spot severity was significantly reduced in the cultivar Dane when compared to MN1410. Roller-crimped rye increased the yield of Dane as compared to full tillage, but there were no differences in yield of MN1410 with cover crop termination treatment.

TABLE 46. *Effect of Fungicide on Foliar Disease Severity at Full Seed (R6) Growth Stage and Yield of Corn*

TREATMENT[z]	FLS[y] %	HARVEST MOISTURE %	TEST WEIGHT LB/BU	YIELD[x] BU/ACRE	
Cover crop termination				Dane	MN1410
Full tillage	0.3	12.8	50.7	29.1 b	53 2
Roller-crimped rye	0.4	12.6	54.9	52.0 a	65 9
Cultivar				p=0.0158	p=0.0566
Dane	0.1 b	12.4 b	50.7 b	—	
MN1410	0.6 a	13.1 a	54.9 a	—	
Fungicide programs and rate					
Nontreated control	0.5	12.5	52.1	52.0	
Endura 70 WDG 8.0 fl oz	0.4	12.5	52.3	49.2	
Double Nickel 55 DWG 2 qt	0.2	12.9	54.3	51.3	
Serifel WP 16 fl oz	0.7	13.0	52.9	50.3	
Actinovate AG 12 oz	0.3	12.4	52.1	49.0	
BotryStop 2 lb	0.2	13.1	53.3	48.8	
P-value *till[v]*	0.6481	0.3946	0.0860	0.0252	
P-value *cultivar*	0.0003	0.0001	0.0001	0.0001	
P-value *fungicide*	0.1618	0.0162	0.4855	0.9379	
P-value *till*cultivar*	0.3741	0.5705	0.0001	0.0188	
P-value *till*fungicide*	0.6915	0.0485	0.4606	0.3631	
P-value *cultivar*fungicide*	0.0612	0.2259	0.2945	0.8294	
P-value *till*cultivar*fungicide*	0.7392	0.1507	0.2673	0.5359	

[z] Fungicide treatments were applied on July 19 at full bloom (R2) growth stage. All plots were inoculated with *S. sclerotiorum* at 1.25 g/foot within the seedbed at planting, and 60 sclerotia per plot were spread between the middle two rows before roller-crimping and after tillage.

[y] Frogeye leaf spot severity was visually assessed as a percentage (0–100%) of symptomatic tissue (lesions) per leaf in the upper canopy on 10 plants per plot on August 26. Values for the 10 plants were averaged before analysis. FLS = frogeye leaf spot.

[x] Yields were adjusted to 13% moisture after harvesting on September 28.

[w] All data were analyzed in SAS 9.4 (SAS Institute, Cary, NC). A generalized linear mixed model analysis of variance was performed using PROC GLIMMIX. Values are least squares means, and values with different letters are significantly different based on least squares means test (α=0.05).

EVALUATION OF SEED TREATMENT FOR MANAGEMENT OF SUDDEN DEATH SYNDROME IN SOYBEAN IN NORTHWESTERN INDIANA, 2021 (SOY21-14.PPAC)

M. T. Brown, S. Shim, and D. E. P. Telenko, Department of Botany and Plant Pathology, Purdue University West Lafayette, IN 47907-2054

SOYBEAN (*GLYCINE MAX* P28T14E AND P25A04X)

Sudden death syndrome, *Fusarium virguliforme*

A trial was established at the Pinney Purdue Agricultural Center (PPAC) in Porter County, Indiana. The experiment was a randomized complete block design with four replications. Plots were 10 feet wide and 30 feet long and consisted of four rows, and the two center rows were used for evaluation. The previous crop was corn. Standard practices for soybean production in Indiana were followed. Soybean varieties P28T14E (susceptible) and P25A04X (resistant) were planted in 30-inch row spacing at a rate of 8 seeds/foot on May 24. *F. virguliforme* inoculum was applied at planting at 1.25 g/foot within the seedbed. Seed treatments were applied on seeds before planting. A foliar application of NanoStress was applied at beginning bloom (R1) growth stage to one of the seed treatment programs. All treatments contained a base treatment except the non-treated control. Sudden death syndrome (SDS) foliar disease ratings were assessed on September 16 at beginning maturity (R7) growth stage. SDS in each plot was rated for disease incidence (DI) and disease severity (DS). Disease incidence refers to the percentage of plants with disease symptoms, and disease severity (DS) was rated using a scale of 1–9 where 1 refers to low disease pressure and 9 refers to premature death of the plant. SDS index was then calculated using the equation DX= (DI x DS)/9. Root rot rating was assessed on August 12 at the R4 (full pod) growth stage by visually assessing dark discoloration on roots. The two center rows of each plot were harvested on September 29, and yields were adjusted to 13% moisture. All data were analyzed in SAS 9.4 (SAS Institute, Cary, NC). A generalized linear mixed model analysis of variance was performed using PROC GLIMMIX. Values are least squares means, and values with different letters are significantly different based on least squares means test (α=0.05).

In 2021, weather conditions were not favorable for disease. SDS was the most prominent disease in the trial but only reached low incidence and severity. There were no significant interactions between cultivar and seed treatments; therefore, main effects are presented. No significant differences were observed in root rot for cultivar or seed treatments on August 12 (Table 47). The resistant cultivar P25A04X had significantly lower levels of SDS index over the susceptible cultivar, P28T14E. No significant differences were observed in percent of canopy green for cultivar or seed treatments on September 16. The resistant cultivar yielded more than the susceptible cultivar, and all the seed treatments resulted in higher yields over the nontreated control.

TABLE 47. *Effect of Cultivar and Seed Treatments on Root Rot, Sudden Death Syndrome (SDS) Index, Canopy Greenness, and Yield of Soybean*

CULTIVAR, TREATMENT AND TIMING[z]	ROOT ROT %[y]	SDS INDEX[w]	CANOPY[w] %	YIELD[v] BU/ACRE
Cultivar				
P25A04X (R)	30.4	0.0 b	32.3	72.5 a
P28T14E (S)	29.1	21.7 a	34.8	66.6 b
Seed treatment programs				
Nontreated control	27.5	14.0	29.4	65.0 b
BASF Base	27.5	8.2	31.3	70.4 a
BASF Base + ILeVO	27.1	12.6	38.8	71.3 a
BASF Base + Saltro	28.6	9.5	26.9	69.1 a
BASF Base + CeraMax	33.7	10.7	38.1	70.1 a
BASF Base + ILeVO fb NanoStress 4 fl oz at R1	31.6	9.5	38.1	71.6 a
Albaugh Base + Mertect 340F + HeadsUp + BioST VPH	31.7	10.9	36.9	70.2 a
Albaugh Base + Mertect 340F + HeadsUp + BioST VPH + ILeVO + TWO.O	30.5	11.3	29.4	63.9 a
P-value *cultivar*[u]	*0.3627*	*0.0001*	*0.3611*	*0.0001*
P-value *seed treatment*	*0.2221*	*0.7233*	*0.1595*	*0.0119*
P-value *cultivar by seed treatment*	*0.3727*	*0.7233*	*0.2453*	*0.5881*

[z] Soybean varieties included SDS susceptible (S) and resistant (R). Seed treatments were applied before planting on May 24. BASF Base contained Allegiance Fl at 4.0 g a/100 kg, Stamina at 7.5 g a/100 kg, Systiva XS Xemium Brand at 5.0 g a/100 kg, Poncho 600 at 0.11 mg a/seed, Flo Rite 1706 at 66.0 ml/100 kg, and Color Coat Red at 33.0 ml/100 kg. Albaugh Base contained Allegiance Fl at 16.0 g a/100 kg, Flo Rite 1706 at 66.0 ml/100 kg, Color Coat Red at 33.0 ml/100 kg, Dynasty at 2.0 g a/100 kg, and Gaucho 600 FS at 0.12 mg a/seed. fb = followed by.

[y] Root rot was visually assessed as a percentage (0–100%) of dark discoloration on roots on August 12.

[x] Disease index was calculated as SDS disease incidence x disease severity (DI x DS)/9.

[w] Canopy green was visually assessed as a percentage (0–100%) of crop canopy green on September 16.

[v] Yields were adjusted to 13% moisture after harvesting on September 29.

[u] All data were analyzed in SAS 9.4 (SAS Institute, Cary, NC). A generalized linear mixed model analysis of variance was performed using PROC GLIMMIX. Values are least squares means, and values with different letters are significantly different based on least squares means test (α=0.05).

COMPARE THE EFFICACY OF SEED TREATMENTS IN SOYBEAN IN NORTHWESTERN INDIANA, 2021 (SOY21-18.PPAC)

S. Shim and D. E. P. Telenko, Department of Botany and Plant Pathology, Purdue University West Lafayette, IN 47907-2054

SOYBEAN (*GLYCINE MAX* P28T14E AND P25A04X)

Sudden death syndrome, *Fusarium virguliforme*

A trial was established at the Pinney Purdue Agricultural Center (PPAC) in Porter County, Indiana. The experiment was a randomized complete block design with four replications. Plots were 10 feet wide and 30 feet long and consisted of four rows, and the two center rows were used for evaluation. The previous crop was corn. Standard practices for soybean production in Indiana were followed. Soybean cultivar P25A04X (resistant) and P28T14E (susceptible) were planted in 30-inch row spacing at a rate of 8 seeds/foot on May 24. Seed treatments were applied on seeds before planting. Disease ratings were assessed on September 8 at full seed (R6) growth stage. Sudden death syndrome (SDS) in each plot was rated for disease incidence (DI) and disease severity (DS). Disease incidence refers to the percentage of plants with disease symptoms, and disease severity (DS) was rated using a scale of 1–9 where 1 refers to low disease pressure and 9 refers to premature death of the plant. SDS index was then calculated using the equation DX= (DI x DS)/9. The two center rows of each plot were harvested on October 1, and yields were adjusted to 13% moisture. All data were analyzed in SAS 9.4 (SAS Institute, Cary, NC). A generalized linear mixed model analysis of variance was performed using PROC GLIMMIX. Values are least squares means, and values with different letters are significantly different based on least squares means test (α=0.05).

In 2021 weather conditions were not favorable for disease, and very little disease developed in plots. SDS was the most prominent disease in the trial. There were no significant differences between seed treatments and root rot rating (Table 48). Resistant cultivar P25A04X had significantly lower levels of SDS incidence, severity, and index over the susceptible cultivar, P28T14E. There were no significant differences between seed treatments and cultivar selection for harvest moisture, test weight and yield.

TABLE 48. *Effect of Seed Treatment on Sudden Death Syndrome (SDS), Root Rot, and Yield of Soybean*

CULTIVAR AND TREATMENT[z]	ROOT ROT[y] %	SDS DI[x]	SDS DS[x]	SDS INDEX[x]	HARVEST MOISTURE %	TEST WEIGHT LB/BU	YIELD[w] BU/ACRE
Nontreated control, P25A04X	9.2	0.3 c	0.3 b	0.0 c	10.0	55.6	78.8
ILeVO	11.6	0.0 c	0.0 b	0.0 c	10.2	56.0	79.9
Saltro	9.7	0.3 c	0.8 b	0.1 c	10.3	56.2	79.7
Nontreated control, P28T14E	14.1	86.3 a	6.0 a	57.5 a	12.4	56.4	73.1
ILeVO	7.0	60.0 b	5.5 a	36.3 b	9.9	54.8	68.7
Saltro	5.1	75.0 ab	5.5 a	45.3 b	9.9	55.4	73.4
P-value[v]	0.1640	0.0001	0.0001	0.0001	0.5329	0.5397	0.0616

[z] Seed treatments were preapplied to the seed of varieties P25A04X (resistant) and P28T14E (susceptible).

[y] Ten roots per plot were sampled from border rows at R4, gently washed, and rated for root rot severity on a scale of 0–100% on August 24.

[x] Sudden death syndrome (SDS) in each plot was rated for disease incidence (DI) and disease severity (DS) on September 8. DI refers to the percentage of plants with disease symptoms, and DS was rated using a scale of 1–9 where 1 refers to low disease pressure and 9 refers to premature death of the plant.

[w] Yields were adjusted to 13% moisture after harvesting on October 1.

[v] All data were analyzed in SAS 9.4 (SAS Institute, Cary, NC). A generalized linear mixed model analysis of variance was performed using PROC GLIMMIX. Values are least squares means, and values with different letters are significantly different based on least squares means test (α=0.05).

EVALUATION OF PLANTING POPULATION, FERTILIZER, AND FUNGICIDE TIMING FOR WHITE MOLD IN SOYBEAN, 2021 (SOY21-22.PPAC)

S. Shim and D. E. P. Telenko, Department of Botany and Plant Pathology, Purdue University West Lafayette, IN 47907-2054

SOYBEAN (*GLYCINE MAX* P34A79X)

White mold, *Sclerotinia sclerotiorum*

A trial was established at the Pinney Purdue Agricultural Center (PPAC) in Porter County, Indiana. The experiment was a randomized complete block design with four replications. Plots were 6.7 feet wide and 30 feet long and consisted of four rows, and the two center rows were used for evaluation. The previous crop was sunflower. Standard practices for soybean production in Indiana were followed. Soybean cultivar P35T15E was planted in 20-inches row spacing at a rate of 8 seeds/foot on May 24. Inoculum of *S. sclerotiorum* was applied on the seedbed at 1.25 g/foot at planting. The field was overhead irrigated weekly at 1 inch unless weekly rainfall was 1 inch or higher to encourage disease. All pesticide applications were applied at 15 gal/acre and 40 psi using a CO_2 backpack sprayer equipped with a 10-foot boom, fitted with four or six TJ-VS 8002 nozzles spaced 20 or 30 inches apart. Fungicides were applied on June 24 at V3 growth stage, July 17 at beginning bloom (R1) growth stage (based on Sporecaster), and July 30 at beginning pod (R3) growth stage. Disease ratings were assessed on September 8 at full seed (R6) growth stage. White mold disease was assessed by counting the number of plants in each plot with symptoms. The two center rows of each plot were harvested on October 1, and yields were adjusted to 13% moisture. All data were analyzed in SAS 9.4 (SAS Institute, Cary, NC). A generalized linear mixed model analysis of variance was performed using PROC GLIMMIX. Values are least squares means, and values with different letters are significantly different based on least squares means test (α=0.05).

In 2021, weather conditions were not favorable for disease. White mold was present in the trial but only remained at low levels. There were no significant differences between fungicide treatments and the non-treated control for disease ratings on September 8 (Table 49). There was no significant effect of treatment on moisture or test weight. Planting at 160,000 seed/acre plus fertilizer resulted in the highest yields as compared to 100,000 seed/acre with no fertilizer.

TABLE 49. *Effect of Fungicide on White Mold Incidence and Yield of Soybean*

TREATMENT, RATE/ACRE, AND TIMING[z]	FERTILIZER	SEEDING RATE SEED/ACRE	WHITE MOLD[y] #/PLOT	HARVEST MOISTURE %	TEST WEIGHT LB/BU	YIELD[x] BU/ACRE
Nontreated control	None	100,000	0.3	10.6	57.4	48.6 e
Endura 70 WDG 8.0 fl oz at R3	None	100,000	0.0	10.4	57.3	50.1 b-e
Endura 70 WDG 8.0 fl oz at Sporecaster	None	100,000	0.0	10.3	57.5	49.4 cde
Cobra 6.0 fl oz at V3	None	100,000	0.0	10.6	56.9	49.1 de
Nontreated control	None	160,000	0.0	10.4	56.7	53.9 a-d
Endura 70 WDG 8.0 fl oz at R3	None	160,000	0.5	10.5	57.0	54.4 abc
Endura 70 WDG 8.0 fl oz at Sporecaster	None	160,000	1.0	10.5	57.0	55.1 ab
Cobra 6.0 fl oz at V3	None	160,000	0.5	10.5	56.8	54.5 abc
Nontreated control	150 lb N	100,000	0.0	10.4	57.3	52.9 a-e
Endura 70 WDG 8.0 fl oz at R3	150 lb N	100,000	0.3	10.6	56.8	51.2 b-e
Endura 70 WDG 8.0 fl oz at Sporecaster	150 lb N	100,000	0.5	10.6	57.2	50.1 b-e
Cobra 6.0 fl oz at V3	150 lb N	100,000	0.0	10.9	57.4	51.0 b-e
Nontreated control	150 lb N	160,000	0.5	10.4	57.1	56.5 a
Endura 70 WDG 8.0 fl oz at R3	150 lb N	160,000	0.3	10.6	56.4	57.6 a
Endura 70 WDG 8.0 fl oz at Sporecaster	150 lb N	160,000	0.3	10.4	57.0	53.6 a-e
Cobra 6.0 fl oz at V3	150 lb N	160,000	0.0	10.5	57.0	50.7 b-e
P-value[w]			0.6431	0.1858	0.0740	0.0200

[z] Fungicide treatments were applied on June 24 at V3 growth stage, July 17 at beginning bloom (R1) growth stage (based on Sporecaster), and July 30 at beginning pod (R3) growth stage. All plots were inoculated with *S. sclerotiorum*.

[y] White mold disease was assessed by counting the number of plants in each plot with symptoms on September 8.

[x] Yields were adjusted to 13% moisture after harvesting on October 1.

[w] All data were analyzed in SAS 9.4 (SAS Institute, Cary, NC). A generalized linear mixed model analysis of variance was performed using PROC GLIMMIX. Values are least squares means, and values with different letters are significantly different based on least squares means test (α=0.05).

EVALUATION OF FUNGICIDES FOR WHITE MOLD IN SOYBEAN IN NORTHWESTERN INDIANA, 2021 (SOY21-25.PPAC)

S. Shim and D. E. P. Telenko, Department of Botany and Plant Pathology, Purdue University West Lafayette, IN 47907-2054

SOYBEAN (*GLYCINE MAX* P35T15E)

White mold, *Sclerotinia sclerotiorum*

A trial was established at the Pinney Purdue Agricultural Center (PPAC) in Porter County, Indiana. The experiment was a randomized complete block design with four replications. Plots were 6.7 feet wide and 30 feet long and consisted of four rows, and the two center rows were used for evaluation. The previous crop was sunflower. Standard practices for soybean production in Indiana were followed. Soybean cultivar P35T15E was planted in 20-inch row spacing at a rate of 8 seeds/foot on May 24. Inoculum of *S. sclerotiorum* was applied on the seedbed at 1.25 g/foot at planting. The field was overhead irrigated weekly at 1 inch unless weekly rainfall was 1 inch or higher to encourage disease. All fungicide applications were applied at 15 gal/acre and 40 psi using a CO_2 backpack sprayer equipped with a 10-foot boom, fitted with four or six TJ-VS 8002 nozzles spaced 20 or 30 inches apart. Fungicides were applied on July 17 at beginning bloom (R1) growth stage and July 30 at beginning pod (R3) growth stage. Disease ratings were assessed on September 8 at full seed (R6) growth stage. White mold disease was assessed by counting the number of plants in each plot with symptoms. The center rows of each plot were harvested on October 1, and yields were adjusted to 13% moisture. All data were analyzed in SAS 9.4 (SAS Institute, Cary, NC). A generalized linear mixed model analysis of variance was performed using PROC GLIMMIX. Values are least squares means, and values with different letters are significantly different based on least squares means test (α=0.05).

In 2021, weather conditions were not favorable for disease. White mold was present in the trial but only remained at low levels. There were no significant differences between fungicide treatments and the non-treated control for disease ratings on September 8 (Table 50). There was no significant effect of treatment on moisture, test weight, and yield of soybean.

TABLE 50. *Effect of Fungicide on White Mold Incidence and Yield of Soybean*

TREATMENT, RATE/ACRE, AND TIMING[z]	WHITE MOLD[y] #/PLOT	HARVEST MOISTURE %	TEST WEIGHT LB/BU	YIELD[x] BU/ACRE
Nontreated control	1.5	11.1	58.0	55.6
Delaro Complete 458 SC 8.0 fl oz at R1	1.8	12.0	56.5	56.5
Delaro Complete 458 SC 8.0 fl oz at R1 fb R3	1.3	11.4	57.2	56.4
Delaro Complete 458 SC 8.0 fl oz at R3	1.0	11.9	56.2	58.8
P-value[w]	0.9317	0.6951	0.2592	0.7590

[z] Fungicide treatments applied on July 17 at beginning bloom (R1) growth stage and July 30 at beginning pod (R3) growth stage, respectively. All plots inoculated with *S. sclerotiorum*. fb = followed by.

[y] White mold disease was assessed by counting the number of plants in each plot with symptoms on September 8.

[x] Yields were adjusted to 13% moisture after harvesting on October 1.

[w] All data were analyzed in SAS 9.4 (SAS Institute, Cary, NC). A generalized linear mixed model analysis of variance was performed using PROC GLIMMIX. Values are least squares means, and values with different letters are significantly different based on least squares means test (α=0.05).

EVALUATION OF FUNGICIDES FOR WHITE MOLD IN SOYBEAN IN NORTHWESTERN INDIANA, 2021 (SOY21-27.PPAC)

S. Shim and D. E. P. Telenko, Department of Botany and Plant Pathology, Purdue University West Lafayette, IN 47907-2054

SOYBEAN (*GLYCINE MAX* P35T15E)

White mold, *Sclerotinia sclerotiorum*

A trial was established at the Pinney Purdue Agricultural Center (PPAC) in Porter County, Indiana. The experiment was a randomized complete block design with four replications. Plots were 6.7 feet wide and 30 feet long and consisted of four rows, and the two center rows were used for evaluation. The previous crop was sunflower. Standard practices for soybean production in Indiana were followed. Soybean cultivar P35T15E was planted in 20-inch row spacing at a rate of 8 seeds/foot on May 24. Inoculum of *S. sclerotiorum* was applied on the seedbed at 1.25 g/foot at planting. The field was overhead irrigated weekly at 1 inch unless weekly rainfall was 1 inch or higher to encourage disease. All fungicide applications were applied at 15 gal/acre and 40 psi using a CO_2 backpack sprayer equipped with a 10-foot boom, fitted with four or six TJ-VS 8002 nozzles spaced 20 or 30 inches apart. Fungicides were applied on July 17 at beginning bloom (R1) growth stage and July 30 at beginning pod (R3) growth stage. Disease ratings were assessed on September 8 at full seed (R6) growth stage. White mold disease was assessed by counting the number of plants in each plot with symptoms. The center rows of each plot were harvested on October 1, and yields were adjusted to 13% moisture. All data were analyzed in SAS 9.4 (SAS Institute, Cary, NC). A generalized linear mixed model analysis of variance was performed using PROC GLIMMIX. Values are least squares means, and values with different letters are significantly different based on least squares means test (α=0.05).

In 2021, weather conditions were not favorable for disease. White mold was present in the trial but only remained at low levels. There were no significant differences between fungicide treatments and the nontreated control for disease ratings on September 8 (Table 51). There was no significant effect of treatment on moisture, test weight, and yield of soybean.

TABLE 51. *Effect of Fungicide on White Mold Incidence and Yield of Soybean*

TREATMENT, RATE/ACRE, AND TIMING[z]	WHITE MOLD[y] #/PLOT	HARVEST MOISTURE %	TEST WEIGHT LB/BU	YIELD[x] BU/ACRE
Nontreated control	1.3	11.7	56.7	49.8
Double Nickel 55 0.5 lb at R1	0.8	11.4	56.5	51.6
Double Nickel 55 0.5 lb at R1 and R3	1.0	11.5	56.9	52.0
LifeGard WG 1.0 oz at R1 and R3	4.0	11.5	57.0	50.2
LifeGard WG 2.0 oz at R1 and R3	2.0	11.4	57.1	52.2
Endura 70 WDG 8.0 oz at R1	0.5	11.5	57.8	53.6
P-value[w]	0.1100	0.9114	0.3530	0.7654

[z] Fungicide treatments were applied on July 17 at beginning bloom (R1) growth stage and on July 30 at beginning pod (R3) growth stage, respectively. All plots were inoculated with *S. sclerotiorum*.

[y] White mold disease was assessed by counting the number of plants in each plot with symptoms September 8.

[x] Yields were adjusted to 13% moisture after harvesting on October 1.

[w] All data were analyzed in SAS 9.4 (SAS Institute, Cary, NC). A generalized linear mixed model analysis of variance was performed using PROC GLIMMIX. Values are least squares means, and values with different letters are significantly different based on least squares means test (α=0.05).

SOUTHWEST PURDUE AGRICULTURAL CENTER (SWPAC)

EVALUATION OF FUNGICIDES FOR FOLIAR DISEASES IN CORN IN SOUTHWESTERN INDIANA, 2021 (COR21-14.SWPAC)

E. A. Duncan, S. Shim, and D. E. P. Telenko, Department of Botany and Plant Pathology, Purdue University West Lafayette, IN 47907-2054

CORN (ZEA MAYS P0574AMXT)

Southern rust, *Puccinia polysora*
Gray leaf spot, *Cercospora zeae-maydis*

A trial was established at the Southwest Purdue Agricultural Center (SWPAC) in Knox County, Indiana. The experiment was a randomized complete block design with four replications. Plots were 10 feet wide and 30 feet long and consisted of four rows, and the two center rows were used for evaluation. The previous crop was soybean. Standard practices for grain corn production in Indiana were followed. Corn hybrid P0574AMXT was planted in 30-inch row spacing at a rate of 27,000 seeds/acre on May 14. All fungicide applications were applied at 15 gal/acre and 40 psi using a CO_2 backpack sprayer equipped with a 10-foot boom, fitted with four TJ-VS 8002 nozzles spaced 30 inches apart. Fungicides were applied on July 13 at silk (R1) growth stage and July 28 at milk (R3) growth stages. Disease rating was assessed on September 9 at maturity (R6) growth stage. Disease severity was rated by visually assessing the percentage of symptomatic leaf area of the ear leaf on five leaves in each plot. Values for each plot were averaged before analysis. The two center rows of each plot were harvested on October 22, and yields were adjusted to 15.5% moisture. All data were analyzed in SAS 9.4 (SAS Institute, Cary, NC). A generalized linear mixed model analysis of variance was performed using PROC GLIMMIX. Values are least squares means, and values with different letters are significantly different based on least squares means test (α=0.05).

In 2021, weather conditions were favorable for disease. Gray leaf spot and southern were the most prominent diseases in the trial and reached moderate severity. All fungicides applied at R1 reduced gray leaf spot over the nontreated controls (Table 52). Headline Amp, Trivapro, Delaro Complete, and Lucento applied at R1 and all fungicides applied at R3 reduced southern rust compared to the nontreated controls. All fungicide applied at R1 and R3 increased percent canopy greenness over the nontreated controls except Headline AMP at R1 and R3. There were no significant differences between treatments for harvest moisture, test weight, and yield of corn.

TABLE 52. *Effect of Fungicide on Foliar Diseases Severity and Yield of Corn*

TREATMENT, RATE/ACRE, AND TIMING[z]	GLS[y] %	SR[y] %	CANOPY[x] %	HARVEST MOISTURE %	TEST WEIGHT LB/BU	YIELD[w] BU/ACRE
Nontreated control 1	8.2 a	6.1 a	37.5 d	14.1	57.4	219.3
Nontreated control 2	5.3 bc	5.9 a	41.3 cd	14.1	57.6	219.7
Headline AMP 1.68 SC 10.0 fl oz at R1	1.8 ef	4.0 bc	50.0 bc	14.1	57.6	228.6
Veltyma 3.34 S 7.0 fl oz at R1	1.4 f	4.9 ab	56.3 ab	14.1	57.6	225.6
Trivapro 2.21 SE 13.7 fl oz at R1	2.3 def	1.4 e	51.3 b	14.1	57.5	221.7
Delaro Complete 458 SC 8.0 fl oz at R1	2.2 def	3.9 bc	51.3 b	14.1	57.3	216.9
Lucento 7.17 SC 5.0 fl oz at R1	1.3 f	0.9 e	52.5 b	14.2	57.4	214.4
Headline AMP 1.68 SC 10 fl oz at R3	6.2 ab	3.9 bc	47.5 bc	14.1	57.5	227.2
Veltyma 3.34 S 7.0 fl oz at R3	4.2 bcd	3.6 bc	56.3 ab	14.3	57.6	216.0
Trivapro 2.21 SE 13.7 fl oz at R3	5.3 bc	1.1 e	62.5 a	14.2	57.6	223.0
Delaro Complete 458 SC 8.0 fl oz at R3	4.1 cd	3.2 cd	48.8 bc	14.1	57.3	214.0
Lucento 7.17 SC 5.0 fl oz at R3	3.8 cde	1.6 de	55.0 ab	14.1	57.5	220.2
P-value[v]	0.0001	0.0001	0.0008	0.7069	0.9998	0.5652

[z] Fungicide treatments were applied on July 13 at silk (R1) growth stage and on July 28 at milk (R3) growth stage. All treatments contained a nonionic surfactant at a rate of 0.25% v/v.

[y] Disease severity was visually assessed as a percentage (0–100%) of symptomatic leaf area on ear leaf, with five plants assessed per plot and ratings averaged before analysis on September 9. GLS = Gray leaf spot, SR = southern rust.

[x] Canopy greenness was visually assessed as a percentage (0–100%) of canopy green on September 9.

[w] Yields were adjusted to 15.5% moisture after harvesting on October 22.

[v] All data were analyzed in SAS 9.4 (SAS Institute, Cary, NC). A generalized linear mixed model analysis of variance was performed using PROC GLIMMIX. Values are least squares means, and values with different letters are significantly different based on least squares means test (α=0.05).

FUNGICIDE COMPARISON FOR FOLIAR DISEASES IN CORN SOUTHWESTERN INDIANA, 2021 (COR21-22.SWPAC)

S. Shim and D. E. P. Telenko, Department of Botany and Plant Pathology, Purdue University West Lafayette, IN 47907-2054

CORN (*ZEA MAYS* P0574AMXT)

Southern rust, *Puccinia polysora*
Gray leaf spot, *Cercospora zeae-maydis*

A trial was established at the Southwest Purdue Agricultural Center (SWPAC) in Knox County, Indiana. The experiment was a randomized complete block design with four replications. Plots were 10 feet wide and 30 feet long and consisted of four rows, and the two center rows were used for evaluation. The previous crop was soybean. Standard practices for grain corn production in Indiana were followed. Corn hybrid P0574AMXT was planted in 30-inches row spacing at a rate of 27,000 seeds/acre on May 14. All fungicide applications were applied at 15 gal/acre and 40 psi using a CO_2 backpack sprayer equipped with a 10-foot boom, fitted with four TJ-VS 8002 nozzles spaced 30 inches apart. Fungicides were applied on July 19 at silk (R1) growth stage. Disease rating was assessed on September 9 at physiological maturity (R6) growth stage. Disease severity was rated by visually assessing as a percentage of symptomatic leaf area of the ear leaf on five leaves in each plot. Values for each plot were averaged before analysis. The two center rows of each plot were harvested on October 22, and yields were adjusted to 15.5% moisture. All data were analyzed in SAS 9.4 (SAS Institute, Cary, NC). A generalized linear mixed model analysis of variance was performed using PROC GLIMMIX. Values are least squares means, and values with different letters are significantly different based on least squares means test (α=0.05).

In 2021, weather conditions were favorable for disease. Gray leaf spot (GLS) and southern rust (SR) were the most prominent diseases in the trial and reached moderate severity. All fungicides application timings significantly reduced SR and GLS compared to the nontreated control on September 9 (Table 53). All fungicide treatments increased the percent if canopy greenness over the nontreated control except Trivapro at 10.3 fl oz and Miravis Neo at 13.7 fl oz. There were no significant differences between treatments for harvest moisture, test weight, and yield of corn.

TABLE 53. *Effect of Fungicide on Foliar Disease Severity and Yield of Corn*

TREATMENT[z]	GLS[y] %	SR[y] %	CANOPY[x] %	HARVEST MOISTURE %	TEST WEIGHT LB/BU	YIELD[w] BU/ACRE
Nontreated control	7.1 a	10.4 a	35.0 c	14.1	57.8	211.6
Quadris 250 SC 4.52 fl oz + Aprovia 5.13 fl oz + Inspire 2.08 SC 5.13 fl oz	3.6 bc	0.5 cd	52.5 ab	14.2	57.8	223.9
Quadris 250 SC 6.02 fl oz + Aprovia 6.84 fl oz+ Inspire 2.08 SC 6.84 fl oz	3.1 c	0.1 d	47.5 ab	14.2	57.8	218.6
Trivapro 2.21 SE 10.3 fl oz	4.0 bc	0.9 cd	43.8 bc	14.2	57.9	210.9
Trivapro 2.21 SE 13.7 fl oz	5.2 ab	0.6 cd	47.5 ab	14.5	58.1	213.9
A23089 325 SC 10.3 fl oz	4.0 bc	2.1 bcb	48.8 ab	14.1	58.0	216.9
A23089 325 SC 13.7 fl oz	3.0 c	2.6 bc	55.0 a	14.2	58.0	220.6
Miravis Neo 2.5 SE 10.3 fl oz	3.5 bc	3.8 b	55.0 a	14.1	58.2	218.5
Miravis Neo 2.5 SE 13.7 fl oz	3.2 c	2.3 bcd	43.8 bc	14.3	57.9	212.5
A23120 340.2 SC 13.7 fl oz	4.5 bc	1.1 cd	47.5 ab	14.2	58.0	217.0
P-value[v]	0.0085	0.0001	0.0319	0.5552	0.9521	0.1361

[z] Fungicide treatments were applied on July 19 at the R1 growth stage, and all treatments contained a nonionic surfactant at a rate of 0.25% v/v.

[y] Disease severity was visually assessed as a percentage (0–100%) of symptomatic leaf area on ear leaf, with five plants assessed per plot and ratings averaged before analysis on September 9. GLS = gray leaf spot, SR=southern rust.

[x] Canopy greenness was visually assessed as a percentage (0–100%) of canopy green on September 9.

[w] Yields were adjusted to 15.5% moisture after harvesting on October 22.

[v] All data were analyzed in SAS 9.4 (SAS Institute, Cary, NC). A generalized linear mixed model analysis of variance was performed using PROC GLIMMIX. Values are least squares means, and values with different letters are significantly different based on least squares means test (α=0.05).

EVALUATION OF FUNGICIDES FOR FOLIAR DISEASES ON SOYBEAN IN SOUTHWESTERN INDIANA, 2021 (SOY21-19.SWPAC)

S. Shim and D. E. P. Telenko, Department of Botany and Plant Pathology, Purdue University West Lafayette, IN 47907-2054

SOYBEAN (*GLYCINE MAX* P35T15E)

Septoria brown spot, *Septoria glycines*
Cercospora leaf blight, *Cercospora kikuchii*

A trial was established at the Southwest Purdue Agricultural Center (SWPAC) in Knox County, Indiana. The experiment was a randomized complete block design with four replications. Plots were 10 feet wide and 30 feet long and consisted of four rows, and the two center rows were used for evaluation. The previous crop was corn. Standard practices for soybean production in Indiana were followed. Soybean cultivar P35T15E was planted in 30-inch row spacing at a rate of 175,000 seed/acre on May 15. All fungicides were applied at 15 gal/acre and 40 psi using a Lee self-propelled sprayer equipped with a 10-foot boom, fitted with six TJ-VS 8002 nozzles spaced 20 inches apart. Fungicides were applied on July 28 at beginning pod (R3) growth stage. The two center rows were harvested on October 19, and yields were adjusted to 13% moisture. All data were analyzed in SAS 9.4 (SAS Institute, Cary, NC). A generalized linear mixed model analysis of variance was performed using PROC GLIMMIX. Values are least squares means, and values with different letters are significantly different based on least squares means test (α=0.05).

In 2020, weather conditions were not favorable for soybean disease; therefore, little to no foliar disease developed. No significant treatment differences were detected for harvest moisture, test weight, and yield of soybean (Table 54).

TABLE 54. *Effect of Fungicide on Yield of Soybean*

TREATMENT, RATE/ACRE, AND TIMING[z]	HARVEST MOISTURE %	TEST WEIGHT LB/BU	YIELD[x] BU/ACRE
Nontreated control	14.2	51.9	89.3
Preemptor 3.22 SC/Fortix 5.0 fl oz at R3	14.3	51.6	92.5
Topguard EQ 4.29 5.0 fl oz at R3	14.0	52.1	94.3
Quadris Top SBX 7.0 fl oz at R3	14.4	51.8	90.1
Lucento 4.17 SC 5.0 fl oz at R3	14.2	52.1	93.0
Miravis Top 1.67 SC 13.7 fl oz at R3	14.5	52.4	93.5
Priaxor Xemium 4.0 fl oz at R3	14.1	52.7	90.2
Trivapro 2.21 SE 13.0 fl oz at R3	14.1	52.3	91.7
Delaro 325 SC 8.0 fl oz at R3	14.4	57.0	92.7
Headline 2.09 SC 10.0 fl oz at R3	14.4	52.5	90.9
Veltyma 3.34 S 7.0 fl oz at R3	14.5	51.4	92.2
Revytek 8.0 fl oz at R3	14.3	51.5	91.3
P-value	*0.1711*	*0.4948*	*0.7237*

[z] Fungicide treatments were applied on July 24 at the R3 growth stage, and all treatments contained a nonionic surfactant (Preference) at a rate of 0.25% v/v.

[x] Yields were adjusted to 13% moisture after harvesting on September 30.

[w] Means followed by the same letter are not significantly different based on least squares means test ($\alpha=0.05$).

EVALUATION OF FOLIAR FUNGICIDES FOR SCAB MANAGEMENT IN SOUTHWESTERN INDIANA, 2021 (WHT21-04.SWPAC)

S. Shim and D. E. P. Telenko, Department of Botany and Plant Pathology, Purdue University West Lafayette, IN 47907-2054

WHEAT (TRITICUM AESTIVUM); P25R40

Fusarium head blight, *Fusarium graminearum*

Plots were established at the Southwest Purdue Agricultural Center (SWPAC) in Knox County, Indiana. The experiment was a randomized complete block design with four replications. Plots were 7.5 feet wide and 20 feet long and consisted of 12 rows spaced 7.5 inches apart, and the center of each plot was used for evaluation. The previous crop was corn. Prior to planting, the field was disked and chisel-plowed on October 7, 2020. Nitrogen (46%) at 50 lb/acre was applied on March 7, 2020. On October 17, 2020, wheat cultivar P25R40 was drilled at 7.5 inches spacing. Harmony Extra at 0.8 oz/acre plus AMS at 2 lb/acre plus NIS at 0.25% v/v was applied on March 25, 2020, for weed management. All fungicide applications were applied at 15 gal/acre and 40 psi using a CO_2 backpack sprayer equipped with a 10-foot boom, fitted with six TJ-VS 8002 nozzles spaced 20 inches apart and directed forward and backward at a 45-degree angle. Fungicides were applied on May 11 and May 27, 2021, at Feekes growth stages 10.5.1 and 10.5.1 + 5 days, respectively. All plots were inoculated with a mixture of isolates of *Fusarium graminearum* endemic to Indiana on May 12. The spore suspension (50,000 spores/ml) was applied at 300 ml/plot with the CO_2 handheld sprayer. Disease ratings were assessed on June 2, 2021. Fusarium head blight (FHB) incidence was measured as the number of infected heads out of 60 plants in each plot and calculated as a percentage. FHB severity was rated by visually assessing the percentage of the infected head. FHB index was calculated as (% FHB incidence multiplied by average FHB severity)/100 per plot. Disease severity on leaves was rated by visually assessing the percentage of symptomatic leaf tissue on five flag leaves per plot for leaf blotch. Values for each plot were averaged before analysis. The eight center rows of each plot were harvested with a Kincaid plot combine on June 22, and yields were adjusted to 13.5% moisture. All data were analyzed in SAS 9.4 (SAS Institute, Cary, NC). A generalized linear mixed model analysis of variance was performed using PROC GLIMMIX. Values are least squares means, and values with different letters are significantly different based on least squares means test ($\alpha=0.05$).

In 2021, weather conditions were not favorable for FHB and leaf blotch diseases. FHB was the most prominent disease, and there was little to no leaf blotch detected. FHB incidence and FHB index were reduced by all fungicides over the nontreated control on June 2 (Table 55). FHB severity was reduced by all fungicides except Prosaro, Sphaerex, Miravis Ace at 10.5.1, and Miravis Ace followed by Folicur. The concentration of deoxynivalenol (DON) was reduced over the nontreated control for all treatments (Table 56). There were no treatment differences in Fusarium damaged kernels (FDK), moisture, test weight, or yield of wheat.

TABLE 55. *Effect of Fungicide on Fusarium Head Blight (FHB) and Foliar Diseases in Wheat*

TREATMENT AND RATE/ACRE[z]	FHB DI[y]	FHB DS[x]	FHB INDEX[w]	LEAF BLOTCH[w]
Nontreated control	82.1 a	5.0 a	4.2 a	0.41
Prosaro 421 SC 6.5 fl oz at 10.5.1	43.8 c	3.0 ab	1.4 bcd	0.01
Caramba 90 EC 13.5 fl oz at 10.5.1	63.8 b	2.6 b	1.8 bc	0.01
Sphaerex (BAS 84000F) 7.3 fl oz at 10.5.1	52.5 bc	4.2 ab	2.1 b	0.00
Miravis Ace 5.2 SC 13.7 fl oz at 10.5.1	20.8 d	3.6 ab	0.7 cd	0.01
Miravis Ace 5.2 SC 13.7 fl oz at 10.5.1+ 5 days	10.4 d	2.2 b	0.2 d	0.01
Miravis Ace 5.2 SC 13.7 fl oz at 10.5.1 fb Prosaro 421 SC 6.5 fl oz at 10.5.1 + 5 days	17.5 d	2.4 b	0.4 d	0.05
Miravis Ace 5.2 SC 13.7 fl oz at 10.5.1 fb Caramba 90 EC 13.5 fl oz 10.5.1 + 5 days	23.3 d	2.1 b	0.5 cd	0.01
Miravis Ace 5.2 SC 13.7 fl oz at 10.5.1 fb Folicur 3.6 F 4.0 fl oz at 10.51 + 5 days	22.5 d	2.8 ab	0.6 cd	0.00
P-value[v]	*0.0001*	*0.1711*	*0.0001*	*0.4832*

[z] Fungicide treatments were applied on May 11 and May 17, 2021, at Feekes growth stages 10.5.1 and 10.5.1 + 5 days, respectively. All treatments contained a nonionic surfactant (Preference) at a rate of 0.125% v/v. All plots were inoculated with *Fusarium graminearum* spore suspension (50,000 spores/ml) after the treatment at Feekes 10.5.1. Spore suspension were applied at 300 ml/plot with handheld sprayer on May 12. fb = followed by.

[y] Fusarium head blight disease incidence (FHB DI) was measured as the number of infected heads out of 60 plants in each plot and calculated as a percentage, and FHB disease severity (FHB DS) was rated by visually assessing the percentage of the infected head on June 2.

[x] FHB index was calculated as (total FHB incidence multiplied by average FHB severity)/100 per plot. FHB = Fusarium head blight.

[w] Disease severity of leaf blotch was rated by visually assessing as a percentage of symptomatic leaf tissue on five flag leaves per plot on June 2.

[v] All data were analyzed in SAS 9.4 (SAS Institute, Cary, NC). A generalized linear mixed model analysis of variance was performed using PROC GLIMMIX. Values are least squares means, and values with different letters are significantly different based on least squares means test (α=0.05).

TABLE 56. *Effect of Fungicide on deoxynivalenol (DON), Fusarium Damaged Kernels (FDK), and Yield in Wheat*

TREATMENT AND RATE/ACRE[z]	DON[y] PPM	FDK[x] %	HARVEST MOISTURE %	TEST WEIGHT LB/BU	YIELD[w] BU/ACRE
Nontreated control	2.9 a	17.0	16.0	16.5	115.3
Prosaro 421 SC 6.5 fl oz at 10.5.1	0.9 b	15.8	15.5	16.5	111.6
Caramba 90 EC 13.5 fl oz at 10.5.1	0.8 bc	15.8	16.7	16.9	119.0
Sphaerex (BAS 84000F) 7.3 fl oz at 10.5.1	0.7 bcd	17.3	16.4	17.0	117.9
Miravis Ace 5.2 SC 13.7 fl oz at 10.5.1	0.6 bcd	12.5	17.0	19.4	117.1
Miravis Ace 5.2 SC 13.7 fl oz at 10.5.1+ 5 days	0.5 bcd	13.3	16.9	18.2	119.0
Miravis Ace 5.2 SC 13.7 fl oz at 10.5.1 fb Prosaro 421 SC 6.5 fl oz at 10.5.1 + 5 days	0.3 d	12.5	17.3	19.1	119.9
Miravis Ace 5.2 SC 13.7 fl oz at 10.5.1 fb Caramba 90 EC 13.5 fl oz 10.5.1 + 5 days	0.4 cd	14.5	17.0	18.1	119.3
Miravis Ace 5.2 SC 13.7 fl oz at 10.5.1 fb Folicur 3.6 F 4.0 fl oz at 10.51 + 5 days	0.4 cd	15.8	16.2	17.2	114.6
P-value[v]	0.0001	0.1546	0.9405	0.0356	0.9908

[z] Fungicide treatments were applied on May 11 and May 17, 2021, at Feekes growth stages 10.5.1 and 10.5.1 + 5 days, respectively. All treatments contained a nonionic surfactant (Preference) at a rate of 0.125% v/v. All plots were inoculated with *Fusarium graminearum* spore suspension (50,000 spores/ml) after the treatment at Feekes 10.5.1. Spore suspension were applied at 300 ml/plot with handheld sprayer on May 12. fb = followed by.

[y] Analysis of the mycotoxin deoxynivalenol (DON) was completed by the University of Minnesota DON Testing Lab on June 22.

[x] Fusarium damaged kernels (FDK) were visually assessed as a percentage of Fusarium damaged heads on June 22.

[w] Yields were adjusted to 13.5% moisture after harvesting on June 22.

[v] All data were analyzed in SAS 9.4 (SAS Institute, Cary, NC). A generalized linear mixed model analysis of variance was performed using PROC GLIMMIX. Values are least squares means, and values with different letters are significantly different based on least squares means test (α=0.05).

EVALUATION OF FOLIAR FUNGICIDES AND VARIETIES FOR SCAB MANAGEMENT IN SOUTHWESTERN INDIANA, 2021 (WHT21-05.SWPAC)

S. Shim and D. E. P. Telenko, Department of Botany and Plant Pathology, Purdue University West Lafayette, IN 47907-2054

WHEAT (TRITICUM AESTIVUM); P25R40 AND P25R61

Fusarium head blight, *Fusarium graminearum*

Plots were established at the Southwest Purdue Agricultural Center (SWPAC) in Knox County, Indiana. The experiment was a strip plot design with four replications. Plots were 7.5 feet wide and 20 feet long and consisted of 12 rows spaced 7.5 inches apart, and the center of each plot was used for evaluation. The previous crop was corn. Prior to planting, the field was disked and chisel-plowed on October 7, 2020. Nitrogen (46%) at 50 lb/acre was applied on March 7, 2020. On October 17, 2020, wheat cultivar P25R40 was drilled at 7.5-inch spacing. Harmony Extra at 0.8 oz/acre plus AMS at 2 lb/acre plus NIS at 0.25% v/v was applied on March 25, 2020, for weed management. All fungicide applications were applied at 15 gal/acre and 40 psi using a CO_2 backpack sprayer equipped with a 10-foot boom, fitted with six TJ-VS 8002 nozzles spaced 20 inches apart and directed forward and backward at a 45-degree angle at 3.0 mph. Fungicides were applied on May 11 and May 17, 2021, at Feekes growth stages 10.5.1 and 10.5.1 + 5 days, respectively. All plots were inoculated with a mixture of isolates of *Fusarium graminearum* endemic to Indiana on May 12. The spore suspension (50,000 spores/ml) was applied at 300 ml/plot with the CO_2 handheld sprayer. Disease ratings were assessed on June 2, 2021. Fusarium head blight (FHB) incidence was measured as the number of infected heads out of 60 plants in each plot and calculated as a percentage. FHB severity was rated by visually assessing the percentage of the infected head. FHB index was calculated as (% FHB incidence multiplied by average FHB severity)/100 per plot. Disease severity on leaves was rated by visually assessing the percentage of symptomatic leaf tissue on five flag leaves per plot for leaf blotch. Values for each plot were averaged before analysis. The eight center rows of each plot were harvested with a Kincaid plot combine on June 22, and yields were adjusted to 13.5% moisture. All data were analyzed in SAS 9.4 (SAS Institute, Cary, NC). A generalized linear mixed model analysis of variance was performed using PROC GLIMMIX. Values are least squares means, and values with different letters are significantly different based on least squares means test (α=0.05).

In 2021, weather conditions were not favorable for FHB and leaf blotch diseases. FHB was the most prominent disease. FHB incidence, severity, and index were reduced by all fungicides over the nontreated, inoculated control in both varieties on June 2 (Table 57). There were no differences detected for leaf blotch. The concentration of deoxynivalenol (DON) was reduced over the nontreated control for all treatments in both varieties (Table 58). Fungicides reduced Fusarium damaged kernels (FDK) in the scab-susceptible cultivar, P25R40, but there were no differences in the resistant cultivar, P25R61. Moisture and test weights were higher in Miravis Ace followed by Folicur in the P25R40 cultivar, and there were no differences between treatments in P25R61. There were no significant differences between treatments in test weight and yield of wheat for either cultivar.

TABLE 57. *Effect of Cultivar and Fungicide on Fusarium Head Blight (FHB) and Foliar Diseases in Wheat*

CULTIVAR, TREATMENT AND RATE/ACRE[z]	FHB DI[y]	FHB DS[y]	FHB INDEX[x]	LEAF BLOTCH %[w]
P25R40				
Nontreated control, inoculated control	89.2 a	5.2 a	4.7 a	0.5
Nontreated, noninoculated control	82.5 c	3.9 b	3.2 b	0.3
Prosaro 421 SC 6.5 fl oz at 10.5.1	58.3 b	2.9 bc	1.6 c	0.5
Miravis Ace 5.2 SC 13.7 fl oz at 10.5.1	34.2 c	2.2 c	0.7 d	1.3
Miravis Ace 13.7 fl oz at 10.5.1 fb Folicur 3.6 F 4.0 fl oz at 10.5.1 + 5 d	32.1 a	1.9 c	0.6 d	1.5
P-value	*0.0001*	*0.0003*	*0.0001*	*0.8225*
P25R61				
Nontreated control, inoculated control	68.8 a	2.1 a	1.4 a	2.0
Nontreated, noninoculated control	54.2 ab	1.8 ab	1.0 ab	2.5
Prosaro 421 SC 6.5 fl oz at 10.5.1	40.8 bc	1.5 bc	0.6 bc	0.3
Miravis Ace 5.2 SC 13.7 fl oz at 10.5.1	19.6 c	1.1 c	0.2 c	0.0
Miravis Ace 13.7 fl oz at 10.5.1 fb Folicur 3.6 F 4.0 fl oz at 10.5.1 + 5 d	17.1 c	1.4 bc	0.3 c	0.0
P-value	*0.0026*	*0.0277*	*0.0051*	*0.0547*

[z] Fungicide treatments were applied on May 11 and May 17 2021, at Feekes growth stages 10.5.1 and 10.5.1 + 6 days, respectively. All treatments contained a nonionic surfactant (Preference) at a rate of 0.125% v/v. All plots were inoculated with *Fusarium graminearum* spore suspension (50,000 spores/ml) after the treatment at Feekes 10.5.1. Spore suspension were applied at 300 ml/plot with a handheld sprayer on May 12. fb = followed by.

[y] Fusarium head blight disease incidence (FHB DI) was measured as the number of infected heads out of 60 plants in each plot and calculated as a percentage. FHB disease severity (FHB DS) was rated by visually assessing the percentage of the infected head.

[x] FHB index was calculated as (FHB DI multiplied by average FHB DS)/100 per plot.

[w] Disease severity of leaf blotch was rated by visually assessing the percentage of symptomatic leaf tissue on five flag leaves per plot.

[v] All data were analyzed in SAS 9.4 (SAS Institute, Cary, NC). A generalized linear mixed model analysis of variance was performed using PROC GLIMMIX. Values are least squares means, and values with different letters are significantly different based on least squares means test (α=0.05).

TABLE 58. *Effect of Fungicide on Deoxynivalenol (DON), Fusarium Damaged Kernels (FDK), and Yield in Wheat*

TREATMENT AND RATE/ACRE[z]	DON[y] PPM	FDK[x] %	HARVEST MOISTURE %	TEST WEIGHT LB/BU	YIELD[w] LB/ACRE
P25R40					
Nontreated control, inoculated control	1.83 a	15.0 ab	16.0 b	55.6 c	96.2
Nontreated, noninoculated control	2.67 a	18.8 a	16.1 b	55.4 c	101.0
Prosaro 421 SC 6.5 fl oz at 10.5.1	0.91 b	10.8 b	16.6 b	56.5 b	104.0
Miravis Ace 5.2 SC 13.7 fl oz at 10.5.1	0.57 b	10.0 b	16.8 b	57.3 a	105.8
Miravis Ace 13.7 fl oz at 10.5.1 fb Folicur 3.6 F 4.0 fl oz at 10.5.1 + 5 d	0.23 b	12.8 b	18.0 a	57.1 ab	108.9
P-value	0.0008	0.0432	0.0030	0.0003	0.5901
P25R61					
Nontreated control, inoculated control	0.52 a	16.5	15.9	54.1 ab	90.8
Nontreated, noninoculated control	0.65 a	15.8	16.5	50.4 b	100.2
Prosaro 421 SC 6.5 fl oz at 10.5.1	0.19 b	15.8	17.6	55.3 a	96.0
Miravis Ace 5.2 SC 13.7 fl oz at 10.5.1	0.10 b	13.8	17.8	55.8 a	97.2
Miravis Ace 13.7 fl oz at 10.5.1 fb Folicur 3.6 F 4.0 fl oz at 10.5.1 + 5 d	0.08 b	12.5	16.9	56.7 a	99.4
P-value	0.0015	0.1551	0.3187	0.0390	0.4993

[z] Fungicide treatments were applied on May 11 and May 17, 2021, at Feekes growth stages 10.5.1 and 10.5.1 + 6 days, respectively. All treatments contained a nonionic surfactant (Preference) at a rate of 0.125% v/v. All plots were inoculated with *Fusarium graminearum* spore suspension (50,000 spores/ml) after the treatment at Feekes 10.5.1. Spore suspension was applied at 300 ml/plot with a handheld sprayer on May 12. fb = followed by.

[y] Analysis of the mycotoxin deoxynivalenol (DON) was completed by the University of Minnesota DON Testing Lab.

[x] Fusarium damaged kernels (FDK) were visually assessed as a percentage of Fusarium damaged heads.

[w] Yields were adjusted to 13.5% moisture after harvesting on June 22.

[v] All data were analyzed in SAS 9.4 (SAS Institute, Cary, NC). A generalized linear mixed model analysis of variance was performed using PROC GLIMMIX. Values are least squares means, and values with different letters are significantly different based on least squares means test (α=0.05).

DAVIS PURDUE AGRICULTURAL CENTER (DPAC)

FIELD-SCALE EVALUATION OF FUNGICIDES FOR FOLIAR DISEASES IN CORN IN CENTRAL INDIANA, 2021 (COR21-09.DPAC)

K. G. Waibel, S. C. Boyer, and D. E. P. Telenko, Department of Botany and Plant Pathology, Purdue University West Lafayette, IN 47907-2054

CORN (*ZEA MAYS* SCS989AM)

Gray leaf spot, *Cercospora zeae-maydis*
Tar spot, *Phyllachora maydis*
Southern rust, *Puccinia polysora*

A trial was established at the Davis Purdue Agricultural Center (DPAC) in Randolph County, Indiana. The experiment was a randomized complete block design with four replications. Plots were 30 feet wide and 500 feet long and consisted of 12 rows, and the two center rows were used for evaluation. The previous crop was soybean. Standard practices for nonirrigated soybean production in Indiana were followed. Corn hybrid P0574AMXT was planted in 30-inch row spacing at a rate of 30,000 seeds/acre on May 21. All fungicide applications were applied at 20 gal/acre and 40 psi using an Apache 720 sprayer. Fungicides were applied on June 28 at V6 growth stage, and on July 21 at tassel/silk (VT/R1) growth stage. Weather conditions prevented a V10 application. Southern rust (SR), tar spot, and gray leaf spot (GLS) were assessed on September 1 at dent (R5) growth stage. Disease severity was rated by visually assessing the percentage of symptomatic leaf area on 10 plants in each plot at the ear leaf on September 1. Ten plants in three locations were assessed in each plot and averaged before analysis. The 12 rows of each plot were harvested on November 16, and yields were adjusted to 15.5 % moisture. All data were analyzed in SAS 9.4 (SAS Institute, Cary, NC). A generalized linear mixed model analysis of variance was performed using PROC GLIMMIX. Values are least squares means, and values with different letters are significantly different based on least squares means test (α=0.05).

In 2021, weather conditions were moderately favorable for disease. GLS was the most prominent disease in the trial and reached low severity. Tar spot and SR were also detected at a low level. The Delaro treatment at V6 application significantly reduced GLS severity over the nontreated control (Table 59). Percent of canopy green was significantly higher in the V6 plots over the VT/R1 application. There was no significant difference between treatments for tar spot and SR severity, harvest moisture, and corn yield.

TABLE 59. *Effect of Fungicide on Foliar Diseases Severity, Canopy Green, and Yield of Corn*

TREATMENT, RATE/ACRE, AND TIMING[z]	GLS[y] %	SR[y] %	CANOPY[x] %	HARVEST MOISTURE %	YIELD[w] BU/ACRE
Nontreated control	2.1 a	0.01	68.8 bc	17.7	207.2
Nonreated control	1.6 a	0.00	75.0 ab	17.8	209.3
Delaro 325 SC 8.0 fl oz at V6	0.6 b	0.04	78.8 a	17.9	212.6
Delaro 325 SC 8.0 fl oz at VT/R1	1.9 a	0.04	67.5 c	17.7	210.0
P-value[v]	0.0083	0.1764	0.0314	0.3694	0.3707

[z] Fungicide treatments were applied on June 28 at V6 growth stage and on July 21 at tassel/silk (VT/R1) growth stage, and all treatments contained a nonionic surfactant (Preference) at a rate of 0.25% v/v.

[y] Disease severity was visually assessed as a percentage (0–100%) of symptomatic leaf area on ear leaf on September 1. Ten leaves were assessed per plot and averaged. SR = southern rust, GLS = gray leaf spot.

[x] Canopy greenness was visually assessed as a percentage (0–100%) of crop canopy green on September 1.

[w] Yields were adjusted to 15.5% moisture after harvesting on November 16.

[v] All data were analyzed in SAS 9.4 (SAS Institute, Cary, NC). A generalized linear mixed model analysis of variance was performed using PROC GLIMMIX. Values are least squares means, and values with different letters are significantly different based on least squares means test (α=0.05).

FIELD-SCALE FUNGICIDE TIMING COMPARISON FOR FOLIAR DISEASES ON SOYBEAN IN CENTRAL INDIANA, 2021 (SOY21-10.DPAC)

K. G. Waibel, J. Boyer, and D. E. P. Telenko, Department of Botany and Plant Pathology, Purdue University West Lafayette, IN 47907-2054

SOYBEAN (*GLYCINE MAX* P35T15E)

Frogeye leaf spot, *Cercospora sojina*
Septoria brown spot, *Septoria glycines*
Downy mildew, *Peronospora manshurica*

A trial was established at the Davis Purdue Agricultural Center (DPAC) in Randolph County, Indiana. The experiment was a randomized complete block design with four replications. Plots were 30 feet wide and 480 feet long and consisted of 24 rows, and the two center rows were used for evaluation. The previous crop was corn. Standard practices for nonirrigated soybean production in Indiana were followed. Soybean cultivar P35T15E was planted in 15-inch row spacing at a rate of 150,000 seeds/acre on May 25. All fungicide applications were applied at 20 gal/acre and 40 psi using Apache 720 sprayer with Trimble CFX monitor. Fungicides were applied on July 27 at beginning pod (R3) growth stage and August 10 at beginning seed (R5) growth stage. Weather conditions prevented a V4 application. Disease ratings were assessed on September 1 at full seed (R6) growth stage. Septoria brown spot (SBS), frogeye leaf spot (FLS), and downy mildew (DM) were rated for disease severity by visually assessing the percentage of symptomatic leaf area in the upper and lower canopies. The soybeans were harvested on November 5, and yields were adjusted to 13% moisture. All data were analyzed in SAS 9.4 (SAS Institute, Cary, NC). A generalized linear mixed model analysis of variance was performed using PROC GLIMMIX. Values are least squares means, and values with different letters are significantly different based on least squares means test (α=0.05).

In 2021, weather conditions were not favorable for disease. SBS and FLS were the most prominent diseases and reached low severity. There was no significant difference between treatments and nontreated controls for disease severity and harvest moisture (Table 60). Delaro applied at R5 had the highest yield but was not significantly different from nontreated control 2 and Delaro applied at R3.

TABLE 60. *Effect of Fungicide on Foliar Disease Severity and Yield of Soybean*

TREATMENT, RATE/ACRE, AND TIMING[z]	FLS[y] UPPER CANOPY %	FLS[y] LOWER CANOPY %	DM[y] %	SBS[y] %	HARVEST MOISTURE %	YIELD[x] EU/ACRE
Nontreated control 1	0.3	0.1	0.2	2.3	14.2	67.6 b
Nontreated control 2	0.6	0.2	0.4	2.9	14.3	70.5 a
Delaro 325 SC 12 fl oz at R3	0.4	0.1	0.4	2.8	14.2	69.6 ab
Delaro 325 SC 12 fl oz at R5	0.5	0.1	0.4	2.5	14.2	71.3 a
P-value[w]	0.6085	0.7778	0.8918	0.2091	0.5935	0.0364

[z] Fungicide treatments were applied on July 27 at beginning pod (R3) growth stage and August 10 at beginning seed (R5) growth stage. Weather conditions prevented a V4 application, and all treatments contained a nonionic surfactant (Preference) at a rate of 0.25% v/v.

[y] Foliar disease severity was visually rated on a scale of 0–100% of upper and lower canopies with disease symptoms on September 1. FLS = frogeye leaf spot, DM = downy mildew, SBS = Septoria brown spot.

[x] Yields were adjusted to 13% moisture after harvesting on November 5.

[w] All data were analyzed in SAS 9.4 (SAS Institute, Cary, NC). A generalized linear mixed model analysis of variance was performed using PROC GLIMMIX. Values are least squares means, and values with different letters are significantly different based on least squares means test (α=0.05).

NORTHEAST PURDUE AGRICULTURAL CENTER (NEPAC)

FIELD-SCALE FUNGICIDE TIMING COMPARISON FOR FOLIAR DISEASES ON CORN IN NORTHEASTERN INDIANA, 2021 (COR21-10.NEPAC)

K. G. Waibel, S. C. Boyer, and D. E. P. Telenko, Department of Botany and Plant Pathology, Purdue University West Lafayette, IN 47907-2054

CORN (ZEA MAYS P0574AMXT)

Gray leaf spot, *Cercospora zeae-maydis*
Tar spot, *Phyllachora maydis*
Southern rust, *Puccinia polysora*

A trial was established at the Northeast Purdue Agricultural Center (NEPAC) in Whitley County, Indiana. The experiment was a randomized complete block design with four replications. Plots were 30 feet wide and 400 feet long and consisted of 12 rows, and the two center rows were used for evaluation. The previous crop was soybean. Standard practices for nonirrigated corn production in Indiana were followed. Corn hybrid P0574AMXT was planted in 30-inch row spacing at a rate of 32,000 seeds/acre on May 16. Fungicide treatments were applied on July 6, July 13, July 21, August 27, and August 3 at the V6, V10, tassel/silk (VT/R1), blister (R2), and milk (R3) growth stages, respectively. Disease ratings were assessed on August 30 at the dent (R5) growth stage. Gray leaf spot (GLS), tar spot, and southern rust (SR) were rated for disease severity by visually assessing the percentage (0–100%) of symptomatic leaf area on the ear leaf on 10 plants at three locations in each plot. The trial was harvested on October 19, and yields were adjusted to 15.5% moisture. All data were analyzed in SAS 9.4 (SAS Institute, Cary, NC). A generalized linear mixed model analysis of variance was performed using PROC GLIMMIX. Values are least squares means, and values with different letters are significantly different based on least squares means test (α=0.05).

In 2021, weather conditions were moderately favorable for disease. GLS, SR, and tar spot were the most prominent diseases in the trial and reached low severity. All Headline Amp application significantly reduced tar spot, SR, and GLS severity over the nontreated control on August 30 except for R3 application on GLS (Table 61). Harvest moisture was significantly higher with all fungicide timings over the nontreated control. There was no significant effect of fungicide timing on yield of corn.

TABLE 61. *Effect of Fungicide on Foliar Diseases Severity and Yield of Corn*

TREATMENT, RATE/ACRE, AND TIMING[z]	TAR SPOT[y] %	SR[y] %	GLS[y] %	HARVEST MOISTURE %	YIELD[x] BU/ACRE
Nontreated control	0.9 a	0.8 a	1.0 a	18.8 c	220.0
Headline AMP 1.68 SC 10.0 fl oz at V6	0.5 b	0.1 b	0.1 c	19.2 b	230.6
Headline AMP 1.68 SC 10.0 fl oz at V10	0.5 b	0.1 b	0.1 c	19.5 a	227.1
Headline AMP 1.68 SC 10.0 fl oz at VT/R1	0.3 b	0.1 b	0.3 c	19.5 a	230.9
Headline AMP 1.68 SC 10.0 fl oz at R2	0.2 b	0.2 b	0.6 b	19.4 ab	226.6
Headline AMP 1.68 SC 10.0 fl oz at R3	0.2 b	0.1 b	0.8 a	19.5 a	225.6
P-value[w]	0.0036	0.0001	0.0001	0.0002	0.1095

[z] Fungicide treatments were applied on July 6, July 13, July 21, August 27, and August 3 at the V6, V10, tassel/silk (VT/R1), blister (R2), and milk (R3) growth stages, respectively.

[y] Disease severity was visually assessed as a percentage (0–100%) of symptomatic leaf area on the ear leaf. Ten leaves were assessed per plot and averaged on August 30. GLS = gray leaf spot, SR = southern rust.

[x] Yields were adjusted to 15.5% moisture after harvesting on October 19.

[w] All data were analyzed in SAS 9.4 (SAS Institute, Cary, NC). A generalized linear mixed model analysis of variance was performed using PROC GLIMMIX. Values are least squares means, and values with different letters are significantly different based on least squares means test (α=0.05).

EVALUATION OF XYWAY 2X2 APPLICATION AT PLANTING FOR FOLIAR DISEASES IN CORN IN NORTHEASTERN INDIANA, 2021 (COR21-19.NEPAC)

K.G. Waibel, S. C. Boyer, and D. E. P. Telenko, Department of Botany and Plant Pathology, Purdue University West Lafayette, IN 47907-2054

CORN (*ZEA MAYS* SCS989AM)

Gray leaf spot, *Cercospora zeae-maydis*
Tar spot, *Phyllachora maydis*
Southern rust, *Puccinia polysora*

A trial was established at the Northeast Purdue Agricultural Center (NEPAC) in Whitley County, Indiana. The experiment was a randomized complete block design with nine replications. Plots were 30 feet wide and 400 feet long and consisted of 12 rows, and the two center rows were used for evaluation. The previous crop was soybean. Standard practices for nonirrigated corn production in Indiana were followed. Corn hybrid SCS89AM was planted in 30-inch row spacing at a rate of 32,000 seeds/acre on May 16. Xyway fungicide was applied with the starter fertilizer with a 2x2 configuration (two inches below and two inches to the side of the seed furrow) in 15 gal/acre with a 12:3 mixture of 28% nitrogen and ammonium thiosulfate at planting. Disease ratings were assessed on August 24 at dent (R_5) growth stage. Gray leaf spot (GLS), tar spot, and southern rust (SR) were rated for disease severity by visually assessing the percentage (0–100%) of symptomatic leaf area on the ear leaf on 10 plants at three locations in each plot. The trial was harvested on October 10, and yields were adjusted to 15.5% moisture. All data were analyzed in SAS 9.4 (SAS Institute, Cary, NC). A generalized linear mixed model analysis of variance was performed using PROC GLIMMIX. Values are least squares means, and values with different letters are significantly different based on least squares means test (α=0.05).

In 2021, weather conditions were not favorable for disease. GLS was the most prominent disease and reached low severity. The 2x2 application of Xyway had significantly lower GLS severity and a higher percent of canopy green compared to the nontreated control (Table 62). There were no significant differences between treatments for tar spot and SR severity and yield of corn.

TABLE 62. *Effect of Fungicide on Foliar Disease Severity and Yield of Corn*

TREATMENT AND RATE/ACRE[z]	GLS[y] %	SR[y] %	TAR SPOT[x] %	CANOPY[w] %	YIELD[v] BU/ACRE
Nontreated control	1.1 a	0.3	0.2	73.8 b	193.0
Xyway LFR 15.2 fl oz 2x2	0.4 b	0.2	0.2	79.3 a	192.9
P-value[u]	0.0001	0.2367	0.3972	0.0120	0.9708

[z] Xyway treatments was applied in starter fertilizer with 2x2-inch spacing from the seed in 15 gal/acre with a 12:3 mixture of 28% nitrogen and ammonium thiosulfate on May 16.

[y] Disease severity was visually assessed as a percentage (0–100%) of symptomatic leaf area on ear leaf on August 24. GLS=gray leaf spot, SR=southern rust.

[x] Tar spot stroma was visually assessed as a percentage (0–100%) of leaf area on ear leaf on August 24.

[w] Canopy greenness was visually assessed as a percentage (0–100%) of canopy green on August 24.

[v] Yields were adjusted to 15.5 % moisture after harvesting on October 10.

[u] All data were analyzed in SAS 9.4 (SAS Institute, Cary, NC). A generalized linear mixed model analysis of variance was performed using PROC GLIMMIX. Values are least squares means, and values with different letters are significantly different based on least squares means test (α=0.05).

FIELD-SCALE FUNGICIDE TIMING FOR FOLIAR DISEASES ON SOYBEAN IN NORTHEASTERN INDIANA, 2021 (SOY21-12.NEPAC)

K. G. Waibel, J. Boyer, and D. E. P. Telenko, Department of Botany and Plant Pathology, Purdue University West Lafayette, IN 47907-2054

SOYBEAN (*GLYCINE MAX* P35T15E)

Frogeye leaf spot, *Cercospora sojina*
Septoria brown spot, *Septoria glycines*
Downy mildew, *Peronospora manshurica*
Sudden death syndrome, *Fusarium virguliforme*

A trial was established at the Northeast Purdue Agricultural Center (NEPAC) in Whitley County, Indiana. The experiment was a randomized complete block design with four replications. Plots were 30 feet wide and 380 feet long. The previous crop was corn. Standard practices for nonirrigated soybean production in Indiana were followed. Soybean cultivar P35T15E was drilled in 7.5-inch row spacing at a rate of 150,000 seeds/acre on May 18. Fungicides were applied at the beginning flower (R1), beginning pod (R3), beginning pod (R5), and R3 followed by (fb) R5 growth stages. Disease ratings were assessed on August 30 at the late R5 growth stage. Septoria brown spot (SBS), frogeye leaf spot (FLS), and downy mildew (DM) were rated for disease severity by visually assessing the percentage of symptomatic leaf area in the upper and lower canopies in three locations in each plot. The soybeans were harvested on October 1, and yields were adjusted to 13% moisture. All data were analyzed in SAS 9.4 (SAS Institute, Cary, NC). All data were analyzed in SAS 9.4 (SAS Institute, Cary, NC). A generalized linear mixed model analysis of variance was performed using PROC GLIMMIX. Values are least squares means, and values with different letters are significantly different based on least squares means test (α=0.05).

In 2021, weather conditions were not favorable for disease. SBS and FLS were the most prominent diseases and reached low severity. All timings of Miravis Top significantly reduced FLS severity in the upper and lower canopies over the nontreated control except Miravis at R3 fb R5 in the lower canopy (Table 63). In addition, a single application of Miravis at R1, R3, and R5 reduced GLS over two applications starting at R3 fb R5. No differences were detected between treatments for DM and SBS severity. Miravis Top applied at R1 increased yield over the nontreated control but was not significantly different from the R5 or R3 fb R5 application timings.

TABLE 63. *Effect of Fungicide Timing on Foliar Disease Severity and Yield of Soybean*

TREATMENT, RATE/ACRE, AND TIMING[z]	FLS[y] UPPER CANOPY %	FLS[y] LOWER CANOPY %	DM[y] %	SBS[y] %	HARVEST MOISTURE %	YIELD[x] BU/ACRE
Nontreated control	2.1 a	0.9 a	0.2	5.8	12.8 a	62.6 bc
Miravis Top 1.67 SC 13.7 oz at R1	0.1 c	0.3 b	0.1	4.2	13.0 a	69.2 a
Miravis Top 1.67 SC 13.7 oz at R3	0.5 c	0.3 b	0.1	5.7	12.2 b	59.2 c
Miravis Top 1.67 SC 13.7 oz at R5	0.3 c	0.3 b	0.2	5.3	12.7 ba	64.3 abc
Miravis Top 1.67 SC 13.7 oz at R3 fb R5	1.4 b	0.7 a	0.3	5.7	13.2 a	66.0 ab
P-value[w]	0.0001	0.0002	0.1953	0.5875	0.0242	0.0260

[z] Fungicide treatments were applied at the beginning flower (R1), beginning pod (R3), beginning pod (R5), and R3 fb R5 growth stages. All treatments contained a nonionic surfactant (Preference) at a rate of 0.25% v/v. fb = followed by.

[y] Foliar disease severity was visually rated on a scale of 0–100% of the upper and lower canopies with disease symptoms August 30. SBS = Septoria brown spot, FLS = Frogeye leaf spot, DM = downy mildew.

[x] Yields were adjusted to 13% moisture after harvesting on October 1.

[w] All data were analyzed in SAS 9.4 (SAS Institute, Cary, NC). A generalized linear mixed model analysis of variance was performed using PROC GLIMMIX. Values are least squares means, and values with different letters are significantly different based on least squares means test (α=0.05).

SOUTHEAST PURDUE AGRICULTURAL CENTER (SEPAC)

FIELD-SCALE EVALUATION OF FUNGICIDE TIMING FOR FOLIAR DISEASES IN CORN IN SOUTHEASTERN INDIANA, 2021 (COR21-10.SEPAC)

K. G. Waibel, J. R. Wahlman, A. Helms, and D. E. P. Telenko, Department of Botany and Plant Pathology, Purdue University West Lafayette, IN 47907-2054

CORN (*ZEA MAYS* P0574AM)

Gray leaf spot, *Cercospora zeae-maydis*

A trial was established at the Southeast Purdue Agricultural Center (SEPAC) in Jennings County, Indiana. The experiment was a randomized complete block design with four replications. Plots were 30 feet wide and 800 feet long and consisted of 12 rows, and the two center rows were used for evaluation. The previous crop was soybean. Standard practices for nonirrigated corn production in Indiana were followed. Corn hybrid P0574AM was planted in 30-inch row spacing at a rate of 29,880 seeds/acre on April 27. All fungicide applications were applied at 20 gal/acre and 40 psi using Apache 720 sprayer. Fungicides were applied on June 17, July 7, and July 20 at V8, V10, and tassel (VT) growth stages, respectively. Disease ratings were assessed on July 28 and August 12 at blister (R2) growth stage and late dough (R4) growth stage, respectively. Gray leaf spot (GLS) was rated for disease severity by visually assessing the percentage (0–100%) of symptomatic leaf area on the ear leaf. Ten plants in three locations were assessed in each plot and averaged before analysis. Twelve rows of each plot were harvested on September 28, and yields were adjusted to 15.5% moisture. All data were analyzed in SAS 9.4 (SAS Institute, Cary, NC). A generalized linear mixed model analysis of variance was performed using PROC GLIMMIX. Values are least squares means, and values with different letters are significantly different based on least squares means test ($\alpha=0.05$).

In 2021, GLS was the most prominent disease and reached moderate severity. On July 28, the V8 and V10 treatments reduced GLS severity over the VT application and the nontreated control (Table 64). All treatments significantly reduced GLS severity over the nontreated control on August 12, with the V10 application having significantly less GLS compared to the V8 and VT applications. No significant differences between treatments were detected for percent of canopy green on September 7. There was no significant difference between treatments for harvest moisture and yield of corn.

TABLE 64. *Effect of Fungicide on Foliar Diseases and Yield of Corn*

TREATMENT, RATE/ACRE, AND TIMING[z]	GLS[y] JUL 28	GLS[y] AUG 12	CANOPY[x] %	HARVEST MOISTURE %	YIELD[w] BU/ACRE
Nontreated control	6.2 a	14.2 a	53.8	14.0	186.5
Lucento 4.17 SC 5.0 fl oz at V8	4.1 b	9.6 b	51.3	14.3	189.7
Lucento 4.17 SC 5.0 fl oz at V10	2.2 c	2.9 c	63.8	14.4	192.3
Lucento 4.17 SC 5.0 fl oz at VT	6.5 a	9.5 b	73.8	14.6	191.3
P-value[v]	0.0009	0.0001	0.2762	0.2762	0.6325

[z] Fungicide treatments were applied on June 17, July 7, and July 20 at V8, V10, and tassel (VT) growth stages, respectively. All treatments contained a nonionic surfactant (Preference) at a rate of 0.25% v/v.

[y] Disease severity was assessed on July 28 and August 12 at blister (R2) growth stage and late dough (R4) growth stages respectively. GLS = gray leaf spot.

[x] Canopy greenness was visually assessed as a percentage (0–100%) of canopy green on September 7.

[w] Yields were adjusted to 15.5% moisture after harvesting on September 28.

[v] All data were analyzed in SAS 9.4 (SAS Institute, Cary, NC). A generalized linear mixed model analysis of variance was performed using PROC GLIMMIX. Values are least squares means, and values with different letters are significantly different based on least squares means test (α=0.05).

FIELD-SCALE FUNGICIDE TIMING COMPARISON FOR FOLIAR DISEASES ON SOYBEAN IN SOUTHEASTERN INDIANA, 2021 (SOY21-11.SEPAC)

K. G. Waibel, J. R. Wahlman, A. Helms, and D. E. P. Telenko, Department of Botany and Plant Pathology, Purdue University West Lafayette, IN 47907-2054

SOYBEAN (*GLYCINE MAX* P34T21SE)

Frogeye leaf spot, *Cercospora sojina*
Septoria brown spot, *Septoria glycines*
Downy mildew, *Peronospora manshurica*

A trial was established at the Southeast Purdue Agricultural Center (SEPAC) in Jennings County, Indiana. The experiment was a randomized complete block design with four replications Plots were 30 feet wide and 600 feet long and consisted of 24 rows, and the two center rows were used for evaluation. The previous crop was corn. Standard practices for nonirrigated soybean production in Indiana were followed. Soybean cultivar P34T21SE was planted in 15-inch row spacing at a rate of 134,000 seeds/acre on April 4. All fungicide applications were applied at 20 gal/acre and 40 psi. Fungicides were applied on June 17, July 20, and August 11 at V4, beginning pod (R3), and beginning seed (R5) growth stages, respectively. Disease ratings were assessed on August 12 at R5 growth stage. Disease severity of each disease was visually assessed as a percentage (0–100%) of symptomatic leaf area in canopy in three locations in each plot on August 12. Frogeye leaf spot (FLS) and downy mildew (DM) were rated in the upper canopy, and Septoria brown spot (SBS) rated in the lower canopy. All ratings were averaged in each plot before analysis. Soybean plots were harvested on September 28, and yields were adjusted to 13% moisture. All data were analyzed in SAS 9.4 (SAS Institute, Cary, NC). A generalized linear mixed model analysis of variance was performed using PROC GLIMMIX. Values are least squares means, and values with different letters are significantly different based on least squares means test (α=0.05).

In 2021 weather conditions were not favorable for disease, and very little disease developed in plots. FLS, DM, and SBS reached low severity. There were no significant differences between treatments for SBS and DM (Table 65). Lucento applied at V4 resulted in the lowest level of SBS in the lower canopy compared to all treatments. No significant differences were observed for soybean yield.

TABLE 65. *Effect of Fungicide on Foliar Disease Severity and Yield of Soybean*

TREATMENT, RATE/ACRE, AND TIMING[z]	FLS[y] %	DM[y] %	SBS[y] %	YIELD[x] BU/ACRE
Nontreated control	0.00	0.1	8.8 a	53.5
Lucento 4.17 SC 5.0 fl oz at V4	0.01	0.1	2.8 b	53.6
Lucento 4.17 SC 5.0 fl oz at R3	0.01	0.2	7.8 a	56.4
Lucento 4.17 SC 5.0 fl oz at R5	0.01	0.1	6.1 a	53.4
P-value[w]	0.8193	0.2259	0.0047	0.0741

[z] Fungicide treatments were applied on June 17, July 20, and August 11 at V4, beginning pod (R3), and beginning seed (R5) growth stages, respectively, and contained a nonionic surfactant (Preference) at a rate of 0.25% v/v.

[y] Foliar disease severity was rated on a scale of 0–100% of canopy with disease symptoms on August 12. Frogeye leaf spot (FLS) and downy mildew (DM) were rated in the upper canopy, and Septoria brown spot (SBS) was rated in the lower canopy.

[x] Yields were adjusted to 13% moisture after harvesting on September 28.

[w] All data were analyzed in SAS 9.4 (SAS Institute, Cary, NC). A generalized linear mixed model analysis of variance was performed using PROC GLIMMIX. Values are least squares means, and values with different letters are significantly different based on least squares means test (α=0.05).

MULTILOCATION TRIALS ON BIOFUNGICIDES

EVALUATION OF THE INTERACTION BETWEEN WHITE MOLD BIOFUNGICIDES AND SYNTHETIC FOLIAR FUNGICIDES IN SOYBEAN IN INDIANA, 2021 (SOY21-05.PPAC AND SOY21-08.ACRE)

A. M. Conrad, S. Shim, and D. E. P. Telenko, Department of Botany and Plant Pathology, Purdue University
West Lafayette, IN 47907-2054

SOYBEAN (*GLYCINE MAX* P34A79X)

Frogeye leaf spot, *Cercospora sojina*
White mold, *Sclerotinia sclerotiorum*

Trials were established at the Agronomy Center for Research and Education (ACRE) in Tippecanoe County, Indiana, and the Pinney Purdue Agricultural Center (PPAC) in Porter County, Indiana. The experiments were a randomized complete block design with four replications. Plots were 6.7 feet wide and 30 feet long and consisted of four rows, and the two center rows were used for evaluation. The previous crop was sunflower. Standard practices for soybean production in Indiana were followed. Soybean cultivar P34A79X was planted in 20-inch row spacing at a rate of 8 seeds/foot on May 15 at ACRE and May 24 at PPAC. All plots were inoculated with *Sclerotinia sclerotiorum* at 1.25 g/foot within the seedbed at planting, and sclerotia at 5 g/plot were spread between the middle two rows prior to emergence. All treatments were applied at 15 gal/acre and 40 psi using a CO_2 backpack sprayer equipped with a 10-foot boom, fitted with six TJ-VS 8002 nozzles spaced 20 inches apart. Contans (*Coniothyrium minitans*) was applied on May 15 at ACRE and on May 26 at PPAC prior to emergence. At ACRE Double Nickel LC (*Bacillus amyloliquefaciens*) was applied on July 13 at full bloom (R2) growth stage. Approach, Endura, and Omega were applied on July 14 at full bloom (R2) growth stage. At PPAC, Double Nickel, Approach, Endura, and Omega were applied on July 19 at full bloom (R2) growth stage. Disease ratings were assessed on September 7 at ACRE and on September 9 at PPAC at full seed (R6) growth stage. Frogeye leaf spot (FLS) severity was rated by visually assessing the percentage (0–100%) of symptomatic tissue per leaf in the upper canopy on 10 plants per plot. Values for the 10 plants were averaged before analysis. Canopy greenness and defoliation were rated on September 14 at ACRE and on September 19 at PPAC. Canopy greenness was visually assessed as a percentage (0–100%) of crop canopy that remained green, and defoliation was visually assessed as a percentage (1–100%) of crop canopy where the leaves had senesced and dropped. The two center rows of each plot were harvested on October 1 at PPAC and October 18 at ACRE, and yields were adjusted to 13% moisture. All data were analyzed in SAS 9.4 (SAS Institute, Cary, NC). A generalized linear mixed model analysis of variance was performed using PROC GLIMMIX. Values are least squares means, and values with different letters are significantly different based on least squares means test (α=0.05).

In 2021 weather conditions were not favorable for disease, and very little disease developed in plots. White mold was not observed in the plots. FLS was the most prominent disease in the trial but only reached low severity. Contans followed by (fb) Endura and Double Nickel fb Endura had the lowest FLS severity but were not statistically different from the nontreated control (Table 66). Contans fb Endura and Contans fb Omega had the highest canopy greenness but were not statistically different from Aproach, Endura, Omega, Contans fb Aproach, Double Nickel fb Aproach, Double Nickel fb Endura, Double Nickel fb Omega, and Contans fb Double Nickle. There was no significant effect of treatment on defoliation, moisture, test weight, or yield of soybean.

TABLE 66. *Effect of Treatment on Frogeye Leaf Spot (FLS) Severity, Canopy Greenness, Defoliation, and Yield of Soybean*

TREATMENT, RATE/ACRE[z]	FLS[y] %	CANOPY[x] %	DEFOLIATION[w] %	MOISTURE %	TEST WEIGHT LBS/BU	YIELD BU/ACRE[v]
Nontreated control	0.6 abc	38.0 bcd	48.6	12.4	54.9	52.7
Contans WG 2.0 lb	0.8 ab	35.6 d	47.5	12.2	54.9	53.3
Double Nickel LC 2.0 qt	0.5 bc	37.5 dc	49.4	12.2	55.1	52.0
Aproach 2.08 SC 12.0 fl oz	0.9 a	39.4 a-d	48.8	12.5	55.1	52.7
Endura 70 WDG 8.0 oz	0.6 abc	38.5 a-d	48.3	12.5	54.6	50.9
Omega 500 F 12.0 fl oz	1.0 a	42.4 abc	44.4	12.4	60.5	53.9
Contans WG 2.0 lb fb Aproach 2.08 SC 12.0 fl oz	0.7 abc	42.5 abc	45.6	12.1	54.9	54.1
Contans WG 2.0 lb fb Endura 70 WDG 8.0 oz	0.4 c	43.8 a	43.8	12.4	54.8	54.3
Contans WG 2.0 lb fb Omega 500 F 12.0 fl oz	0.7 abc	43.8 a	41.9	12.3	55.1	54.0
Double Nickel LC 2.0 qt fb Aproach 2.08 SC 12.0 fl oz	0.7 abc	40.4 a-d	48.4	12.1	55.0	52.3
Double Nickel LC 2.0 qt fb Endura 70 WDG 8.0 oz	0.4 c	39.4 a-d	48.1	12.7	55.1	48.2
Double Nickel LC 2.0 qt fb Omega 500 F 12.0 fl oz	0.6 abc	42.7 abc	44.3	12.4	55.0	53.1
Contans WG 2.0 lb fb Double Nickel LC 2.0 qt	0.7 abc	41.9 abc	45.1	12.2	54.7	52.0
P-value	*0.0213*	*0.0199*	*0.0533*	*0.2905*	*0.3143*	*0.9283*

[z] Contans (*Coniothyrium minitans*) was applied on May 15 at ACRE and on May 26 at PPAC prior to emergence. At ACRE, Double Nickel LC (*Bacillus amyloliquefaciens*) was applied on July 13 at full bloom (R2) growth stage, and Approach, Endura, and Omega were applied on July 14 at full bloom (R2) growth stage. At PPAC, Double Nickel, Approach, En dura, and Omega were applied on July 19 at full bloom (R2) growth stage. Disease ratings were assessed on September 7 at ACRE and on September 9 at PPAC at full seed (R6) growth stage. All plots were inoculated with *S. sclerotiorum* at 1.25 g/foot within the seedbed at planting. fb = followed by.

[y] FLS severity was visually assessed as a percentage (1–100%) of symptomatic tissue (lesions) per leaf in the upper canopy on 10 plants per plot. Values for the 10 plants were averaged before analysis on September 7 at ACRE and on September 9 at PPAC.

[x] Canopy greenness was visually assessed as a percentage (1–100%) of crop canopy still green on September 14 at ACRE and on September 19 at PPAC.

[w] Defoliation was visually assessed as a percentage (1–100%) of crop canopy where the leaves had senesced and dropped.

[v] Yields were adjusted to 13% moisture after harvesting on October 1 at PPAC and on October 18 at ACRE..

[v] All data were analyzed using PROC GLIMMIX in SAS 9.4 (SAS Institute, Cary, NC). Means followed by the same letter are not significantly different based on least squares means test (α=0.05).

EVALUATION OF THE INTERACTION BETWEEN WHITE MOLD BIOFUNGICIDES AND POSTEMERGENCE HERBICIDES IN SOYBEAN IN INDIANA, 2021 (SOY21-04.PPAC AND SOY21-07.ACRE)

A. M. Conrad, S. Shim, and D. E. P. Telenko, Department of Botany and Plant Pathology, Purdue University West Lafayette, IN 47907-2054

SOYBEAN (*GLYCINE MAX* P34A79X)

Frogeye leaf spot, *Cercospora sojina*
White mold, *Sclerotinia sclerotiorum*

Trials were established at the Agronomy Center for Research and Education (ACRE) in Tippecanoe County, Indiana, and the Pinney Purdue Agricultural Center (PPAC) in Porter County, Indiana. The experiments were a randomized complete block design with four replications. Plots were 6.7 feet wide and 30 feet long and consisted of four rows, and the two center rows were used for evaluation. The previous crops were sunflower at ACRE and soybean at PPAC. Soybean cultivar P34A79X was planted in 20-inch row spacing at a rate of 8 seeds/foot on May 15 at ACRE and on May 25 at PPAC. All plots were inoculated with *Sclerotinia sclerotiorum* at 1.25 g/foot within the seedbed at planting. Contans and Double Nickel were applied at 15 gal/acre and 40 psi, and FirstRate and RoundUp PowerMax were applied at 20 gal/acre and 40 psi, in both cases using a CO_2 backpack sprayer equipped with a 10-foot boom, fitted with six TJ-VS 8002 nozzles spaced 20 inches apart. XtendiMax was applied at 20 gal/acre and 30 psi using a CO_2 backpack sprayer equipped with a 10-foot boom, fitted with six TTI 11003 nozzles spaced 20 inches apart. Contans was applied on May 15 at ACRE and on May 26 at PPAC prior to emergence. FirstRate, RoundUp PowerMax, and XtendiMax were applied on June 16 at ACRE and on June 20 at PPAC at second vegetative (V2) growth stage. Double Nickel was applied on July 13 at ACRE at full bloom (R2) growth stage and on July 30 at PPAC at beginning pod (R3) growth stage. Disease ratings were assessed on August 23, September 1, and September 7 at ACRE and on August 26, September 2, and September 9 at PPAC at full seed (R6) growth stage. Frogeye leaf spot (FLS) severity was rated by visually assessing the percentage (0–100%) of symptomatic tissue per leaf in the upper canopy on 10 plants per plot. Values for the 10 plants were averaged before analysis. Canopy greenness and defoliation were rated on September 14 at ACRE and on September 19 at PPAC. Canopy greenness was visually assessed as a percentage (0–100%) of crop canopy that remained green, and defoliation was visually assessed as a percentage (1–100%) of crop canopy where the leaves had senesced and dropped. The two center rows of each plot were harvested on October 18 at ACRE and on September 29 at PPAC, and yields were adjusted to 13% moisture. All data were analyzed in SAS 9.4 (SAS Institute, Cary, NC). A generalized linear mixed model analysis of variance was performed using PROC GLIMMIX. Values are least squares means, and values with different letters are significantly different based on least squares means test (α=0.05).

In 2021 weather conditions were not favorable for disease development, and very little disease developed in plots. White mold was not observed in the plots. FLS was the most prominent disease in the trial but only reached low severity. There were no significant differences between treatments when compared to the nontreated control for FLS severity and defoliation (Table 67). XtendiMax had the highest canopy greenness but was not statistically different from the nontreated control. There was no significant effect of treatment on moisture or test weight. Soybean yield was highest in the Contans followed by (fb) RoundUp Power-Max and RoundUp PowerMax fb Double Nickel treatments but were not statistically different from the FirstRate, RoundUp PowerMax, XtendiMax, Contans fb FirstRate, FirstRate fb Double Nickel, or XtendiMax fb Double Nickel treatments.

TABLE 67. *Effect of Treatment on Frogeye Leaf Spot (FLS) Severity, Canopy Greenness, Defoliation, and Yield of Soybean*

TREATMENT AND RATE/ACRE[z]	FLS[y] %	CANOPY[x] %	DEFOLIATION[w] %	MOISTURE %	TEST WEIGHT LBS/BU	YIELD[v] BU/ACRE
Nontreated control	1.2	40.3 abc	51.2	12.9	54.7	43.5 d
Contans WG 2.0 lbs	0.6	41.3 ab	48.2	13.4	54.9	44.9 dc
Double Nickel LC 2.0 qt	1.0	36.6 bc	48.0	13.3	55.0	45.6 dc
FirstRate WG 0.6 oz	0.7	39.9 abc	43.8	13.1	54.8	52.4 ab
RoundUp PowerMax EC 22.0 fl oz	0.8	38.2 abc	45.8	13.3	54.9	50.8 abc
XtendiMax EC 22.0 fl oz	1.1	43.2 a	45.5	13.0	54.8	48.1 a-d
Contans WG 2.0 lbs fb FirstRate WG 0.6 oz	1.1	35.6 c	45.6	13.2	55.1	48.8 a-d
Contans WG 2.0 lbs fb RoundUp PowerMax EC 22.0 fl oz	1.0	36.2 bc	44.2	13.1	55.3	54.0 a
Contans WG 2.0 lbs fb XtendiMax EC 22.0 fl oz	1.1	40.7 ab	43.8	13.3	55.2	46.6 bcd
FirstRate WG 0.6 oz fb Double Nickel LC 2.0 qt	0.7	35.7 c	44.7	13.2	55.0	52.2 ab
RoundUp PowerMax EC 22.0 fl oz fb Double Nickel LC 2.0 qt	1.2	37.1 cb	44.4	12.9	54.4	54.0 a
XtendiMax EC 22.0 fl oz fb Double Nickel LC 2.0 qt	0.8	34.5 c	50.4	12.9	54.6	49.8 a-d
P-value[u]	*0.1338*	*0.0312*	*0.2752*	*0.8826*	*0.7324*	*0.0295*

[z] Contans was applied on May 15 at ACRE and on May 26 at PPAC prior to emergence. FirstRate, RoundUp PowerMax, and XtendiMax were applied on June 16 at ACRE and on June 20 at PPAC at second vegetative (V2) growth stage, and Double Nickel was applied on July 13 at ACRE at full bloom (R2) growth stage and on July 30 at PPAC at beginning pod (R3) growth stage. All plots were inoculated with *S. sclerotiorum* at 1.25 g/foot within the seedbed at planting. fb = followed by.

[y] Frogeye leaf spot (FLS) severity was rated by visually assessing the percentage (1–100%) of symptomatic tissue (lesions) per leaf in the upper canopy on 10 plants per plot on August 23, September 1, and September 7 at ACRE and August 26, September 2, and September 9 at PPAC at full seed (R6) growth stage. Values for the 10 plants were averaged before analysis.

[x] Canopy greenness was visually assessed as a percentage (1–100%) of crop canopy that stayed green on September 14 at ACRE and on September 19 at PPAC.

[w] Defoliation was visually assessed as a percentage (1–100%) of crop canopy where the leaves had senesced and dropped.

[v] Yields were adjusted to 13% moisture after harvesting on October 18 at ACRE and on September 29 at PPAC.

[u] All data were analyzed using PROC GLIMMIX in SAS 9.4 (SAS Institute, Cary, NC). Means followed by the same letter are not significantly different based on least squares means test (α=0.05).

EVALUATION OF THE INTERACTION BETWEEN WHITE MOLD BIOFUNGICIDES AND PREEMERGENCE HERBICIDES IN SOYBEAN IN INDIANA, 2021 (SOY21-03.PPAC AND SOY21-06.ACRE)

A. M. Conrad, S. Shim, and D. E. P. Telenko, Department of Botany and Plant Pathology, Purdue University West Lafayette, IN 47907-2054

SOYBEAN (*GLYCINE MAX* P34A79X)

Frogeye leaf spot, *Cercospora sojina*
White mold, *Sclerotinia sclerotiorum*

Trials were established at the Agronomy Center for Research and Education (ACRE) in Tippecanoe County, Indiana, and the Pinney Purdue Agricultural Center (PPAC) in Porter County, Indiana. The experiments were a randomized complete block design with four replications. Plots were 6.7 feet wide and 30 feet long and consisted of four rows, and the two center rows were used for evaluation. The previous crops were sunflower at ACRE and soybean at PPAC. Soybean cultivar P34A79X was planted in 20-inch row spacing at a rate of 8 seeds/foot on May 15 at ACRE and on May 25 at PPAC. All plots were inoculated with *Sclerotinia sclerotiorum* at 1.25 g/foot within the seedbed at planting. Contans and Double Nickel were applied at 15 gal/acre and 40 psi, and Valor, Dual Magnum, and Metribuzin were applied at 20 gal/acre and 40 psi, in both cases using a CO_2 backpack sprayer equipped with a 10-foot boom, fitted with six TJ-VS 8002 nozzles spaced 20 inches apart. Contans, Valor, Dual Magnum, and Metribuzin were applied on May 15 at ACRE and on May 26 at PPAC prior to emergence. Double Nickel was applied on July 13 at ACRE at full bloom (R2) growth stage and on July 30 at PPAC at beginning pod (R3) growth stage. Disease ratings were assessed on September 7 at ACRE and on September 9 at PPAC at full seed (R6) growth stage. Frogeye leaf spot (FLS) severity was visually assessed as a percentage (0–100%) of symptomatic tissue per leaf in the upper canopy on 10 plants per plot. Values for the 10 plants were averaged before analysis. Canopy greenness and defoliation were rated on September 14 at ACRE and on September 19 at PPAC. Canopy greenness was visually assessed as a percentage (0–100%) of crop canopy that remained green, and defoliation was visually assessed as a percentage (1–100%) of crop canopy where the leaves had senesced and dropped. The two center rows of each plot were harvested on October 10 at ACRE and on September 29 at PPAC, and yields were adjusted to 13% moisture. All data were analyzed in SAS 9.4 (SAS Institute, Cary, NC). A generalized linear mixed model analysis of variance was performed using PROC GLIMMIX. Values are least squares means, and values with different letters are significantly different based on least squares means test ($\alpha=0.05$).

In 2021 weather conditions were not favorable for disease development, and very little disease developed in plots. White mold was not observed in the plots. FLS was the most prominent disease in the trials but only reached low severity. There were no significant differences between treatments when compared to the non-treated control for FLS severity, canopy greenness, and defoliation (Table 68). There was no significant effect of treatment on moisture, test weight, and yield of soybean.

TABLE 68. *Effect of Treatment on Disease, Canopy Greenness, Defoliation, and Yield of Soybean*

TREATMENT AND RATE/ACRE[z]	FLS[y] %	CANOPY[x] %	DEFOLIATION[w] %	MOISTURE %	TEST WEIGHT LBS/BU	YIELD[v] BU/ACRE
Nontreated control	0.60	41.3	45.9	12.9	55.5	48.7
Contans WG 2.0 lb	0.87	41.9	47.3	12.5	55.1	50.2
Double Nickel LC 2.0 qt	0.92	41.9	50.6	12.4	55.4	50.8
Valor WG 3.0 fl oz	0.90	40.6	51.3	12.4	55.4	48.8
Dual Magnum EC 2.6 pt	0.72	44.4	47.5	12.4	55.3	50.3
Metribuzin DF 1.0 pt	0.96	40.8	47.2	12.4	55.5	52.6
Contans WG 2.0 lb fb Valor WG 3.0 fl oz	0.85	44.4	48.1	12.9	55.3	52.1
Contans WG 2.0 lb fb Dual Magnum EC 2.6 pt	0.55	41.7	46.1	12.6	55.4	48.8
Contans WG 2.0 lb fb Metribuzin DF 1.0 pt	0.58	41.9	48.8	12.4	55.4	49.2
Valor WG 3.0 fl oz fb Double Nickel LC 2.0 qt	0.81	44.4	46.9	12.6	55.5	52.8
Dual Magnum EC 2.6 pt fb Double Nickel LC 2.0 qt	0.85	42.5	47.5	12.5	55.4	50.0
Metribuzin DF 1.0 pt fb Double Nickel LC 2.0 qt	0.94	42.5	48.1	12.6	55.5	53.2
P-value[u]	0.4808	0.9380	0.9637	0.1305	0.9467	0.2390

[z] Contans, Valor, Dual Magnum, and Metribuzin were applied on May 15 at ACRE and on May 26 at PPAC prior to emergence. Double Nickel was applied on July 13 at ACRE at full bloom (R2) growth stage and on July 30 at PPAC at beginning pod (R3) growth stage. All plots were inoculated with *S. sclerotiorum* at 1.25 g/foot within the seedbed at planting. fb = followed by.

[y] Frogeye leaf spot (FLS) severity was visually assessed as a percentage (1–100%) of symptomatic tissue (lesions) per leaf in the upper canopy on 10 plants per plot on September 7 at ACRE and on September 9 at PPAC at full seed (R6) growth stage. Values for the 10 plants were averaged before analysis.

[x] Canopy greenness was visually assessed as a percentage (1–100%) of crop canopy that stayed green on September 14 at ACRE and on September 19 at PPAC.

[w] Defoliation was visually assessed as a percentage (1–100%) of crop canopy where the leaves had senesced and dropped.

[v] Yields were adjusted to 13% moisture after harvesting on October 10 at ACRE and on September 29 at PPAC.

[u] All data were analyzed using PROC GLIMMIX in SAS 9.4 (SAS Institute, Cary, NC). Means followed by the same letter are not significantly different based on least squares means test (α=0.05).

APPENDIX: WEATHER DATA

TABLE 69. *Average Monthly Weather Conditions at the Purdue Agronomy Center for Research and Education (ACRE), Pinney Purdue Agricultural Center (PPAC), Southwest Purdue Agricultural Center (SWPAC), Davis Purdue Agricultural Center (DPAC), Northeast Purdue Agricultural Center (NEPAC), and Southeast Purdue Agricultural Center (SEPAC) in Indiana, 2021[z]*

MONTHS	ACRE			PPAC			SWPAC		
	TEMP. MIN.[y] °F	TEMP. MAX.[y] °F	TOTAL PRECIPIT.[x] (IN)	TEMP. MIN.[y] °F	TEMP. MAX.[y] °F	TOTAL PRECIPIT.[x] (IN)	TEMP. MIN.[y] °F	TEMP. MAX.[y] °F	TOTAL PRECIPIT.[x] (IN)
January	35.5	23.3	1.95	32.3	21.7	1.58	40.9	26.8	2.26
February	30.7	13.1	0.62	25.9	10.1	0.58	36.6	19.4	2.60
March	57.5	33.9	3.53	52.2	30.6	1.49	62.5	39.5	3.55
April	64.1	41.4	2.65	59.6	37.3	1.39	68.4	45.7	2.80
May	71.7	49.3	6.04	67.6	46.5	3.92	76.6	52.4	3.15
June	85.2	63.5	6.21	80.7	60.9	5.63	87.4	65.9	4.50
July	82.8	63.8	4.12	79.3	61.7	2.99	87.0	67.6	5.57
August	85.9	64.3	2.07	82.8	61.6	4.03	88.2	68.0	3.08
September	81.4	57.2	2.12	78.5	54.1	1.23	83.9	61.0	2.65
October	68.3	51.2	8.41	65.6	47.8	6.28	72.6	54.4	5.97
November	48.7	30.8	1.35	46.4	28.7	0.84	53.5	33.0	1.60
December	49.1	30.7	3.18	44.3	26.8	2.52	55.4	36.4	4.11

MONTHS	DPAC			NEPAC			SEPAC		
	TEMP. MIN.[y] °F	TEMP. MAX.[y] °F	TOTAL PRECIP.[x] (IN)	TEMP. MIN.[y] °F	TEMP. MAX.[y] °F	TOTAL PRECIP.[x] (IN)	TEMP. MIN.[y] °F	TEMP. MAX.[y] °F	TOTAL PRECIP.[x] (IN)
January	34.9	23.7	1.68	33.4	23.9	1.13	38.6	26.4	3.15
February	30.5	13.5	1.18	29.6	13.4	0.44	35.9	20.7	2.97
March	56.3	30.9	2.21	55.3	31.0	3.72	59.8	34.9	3.74
April	61.4	37.4	1.34	59.9	37.4	2.04	65.2	42.2	4.01
May	69.9	46.4	3.88	69.5	48.7	5.80	73.2	49.7	2.73
June	82.5	62.7	2.54	82.9	63.0	6.24	83.1	63.9	6.73
July	82.5	63.0	4.02	81.9	63.4	3.45	84.9	64.1	7.13
August	85.6	63.0	2.69	84.4	63.6	4.96	87.8	64.8	2.19
September	78.9	55.1	4.08	79.3	56.6	3.87	81.3	57.1	4.79
October	68.9	49.9	5.40	66.9	50.7	8.90	71.3	51.2	4.79
November	48.9	29.5	1.36	47.3	30.9	1.39	52.8	30.4	2.18
December	49.0	29.9	4.48	45.7	29.4	3.95	54.9	33.5	4.69

[z] Data courtesy of Indiana State Climate Office. Taken from the Purdue Mesonet Data Hub at https://ag.purdue.edu/indiana-state-climate/purdue-mesonet/purdue-mesonet-data-hub/.

[y] Average minimum and maximum temperatures for each month.

[x] Total precipitation for each month.

ABOUT THE AUTHORS

DARCY E. P. TELENKO is an associate professor and Extension plant pathologist in the Department of Botany and Plant Pathology at Purdue University. Her interdisciplinary research and Extension program are involved in studying the biology and management of soilborne and foliar pathogens of agronomic crops. Telenko is a native of western New York and received her PhD at North Carolina State University. She has published more than sixty peer-review manuscripts and two hundred Extension publications. She was awarded the 2024 Leadership Award from the Purdue University Cooperative Extension Specialist Association.

SUJOUNG SHIM is a research associate in the Department of Botany and Plant Pathology at Purdue University. Her research involves designing, conducting, analyzing, and reporting on a variety of research projects. She has a BS in pharmaceutical science and an MS in public health, both from Purdue University. Shim has served as a coauthor on more than ten peer-reviewed publications and twenty-five peer-reviewed technical reports.

www.ingramcontent.com/pod-product-compliance
Lightning Source LLC
Chambersburg PA
CBHW080558220326

41599CB00032B/6527